AF853362

A HISTORY OF THE WOODBURYTYPE

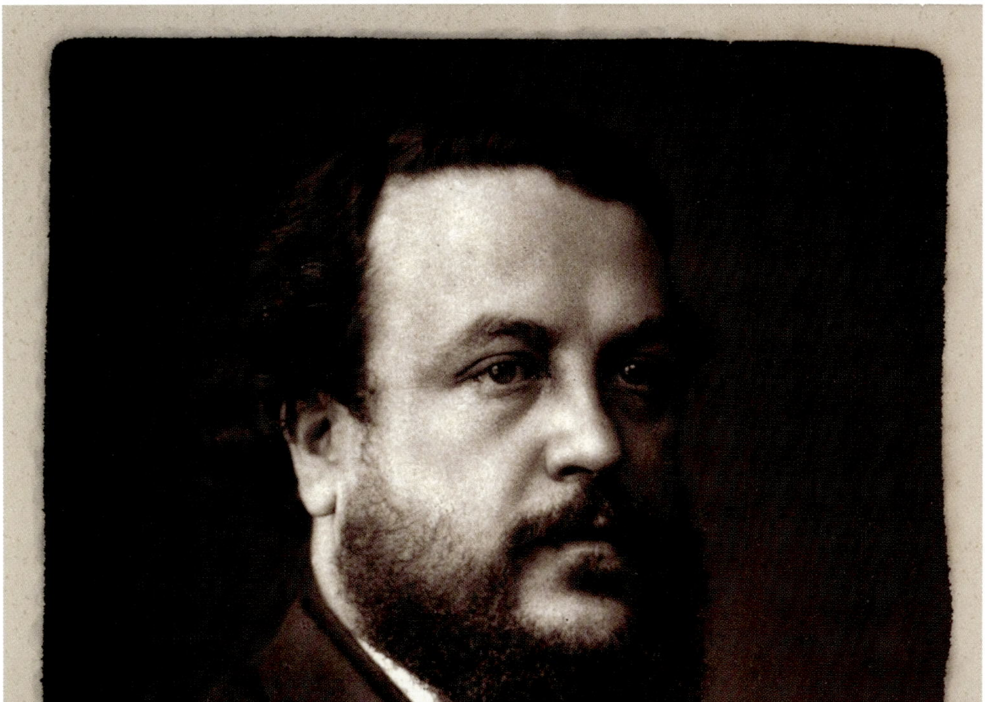

[Self] portrait, Walter Bentley Woodbury, ca. 1880

A HISTORY OF THE
WOODBURYTYPE

The First Successful

PHOTOMECHANICAL PRINTING PROCESS

and Walter Bentley Woodbury

by BARRET OLIVER

CARL MAUTZ PULISHING
NEVADA CITY, CALIFORNIA 2006

First Edition (Limited to 800 copies)
Copyright 2007 Barret Oliver

Cataloging-in-Publication Data

Oliver, Barret
A History of the Woodburytype: the First Successful Photomechanical Printing Process and Walter Bentley Woodbury / Oliver, Barret
p. cm.
Includes bibliographical references and index
ISBN-10: 1-887694-28-5
ISBN-13: 978-1-887694-28-5

1. Photography – History – 19th century. I. Oliver, Barret II. Title

770'.9

Carl Mautz Publishing
329 Bridge Way
Nevada City, California 95959
530-478-1610 Fax 530-478-0772
www.carlmautz.com

DEDICATED TO

Luanne Woodbury

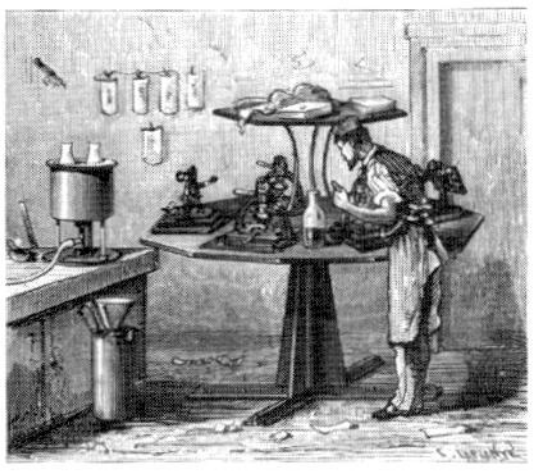

Every time a photographer solves an aesthetic problem he has to solve an underlying technical problem. One solution supports and influences the other. Granted, the photographer is infinitely more interesting than the photographic machine, and so are the things the photographer uses the machine to capture. But unless you understand the ways the machine cajoles and begrudges, you will never know how truly interesting the photographer is. You will not see the compromises he had to make in order to put his vision on paper.

—William Crawford, *Keepers of Light, 1979*

CONTENTS

FOREWORD

What is actual is actual only for one time and only for one place.
—Ezra Pound

Questions concerning image permanence and the almost insatiable demands of society for the creation and mass production of book, journal and newspaper illustrations led to the emergence of a profusion of photomechanical and reprographic processes; more often than not, bearing strange and exotic names. Further examination reveals that all, if not most, were variations on three principal technologies—intaglio (*en*taglio: printing from a raised surface), relief, and planographic, including collotype and other hybrid processes.

Negre's Heliogravure (1865), and Talbot's Photoglyphic Engraving (1852, 1858 and 1866) processes were precursors of Klic's photogravure. However, when Walter Woodbury's process appeared it set a qualitative standard for the reproduction and replication of photographic images and engravings by which all other processes would inevitably be compared and found wanting. It was not until the 1980s, with the advent of drum scanning and digital image processing, that the Woodburytype process was equaled or matched. In addition to which, the final image, being non-photographic, possessed a much higher degree of permanence.

Woodburytype is a strange and exotic hybrid, and, like several other emergent processes, was founded upon the properties of dichromated gelatin, which when exposed to light, created an image, in relief that directly related to the tonal range of the original.[i] This relief, in turn, was used as a matrix to form a second (lead) relief from which an image was cast, in pigmented gelatin, on paper. The resultant image in true continuous tone was screenless, and, to the non-expert, indistinguishable from the most pristine albumen or silver-halide photographic print. By the 1870s, according to Baden Prichard, something on the order of 50,000 prints a day were being produced by the Woodbury Permanent Printing Company in Ealing and matched by the output of Goupil's impressive works at Asnieres, Paris. The process was used extensively for book illustration covering a wide range of subjects, primarily because of its ability to reproduce images with a degree of accuracy and sharpness that was unparalleled. Although Carbon Transfer and Woodburytype processes are often confused and misidentified, the latter displayed a higher degree of acutance, a clearly defined and contoured relief combined with sharp linear detailing along edges of high contrast. Without the use and integration of two contemporaneous technological developments, Tangye's hydraulic press and Swan's Carbon Transfer process, Woodbury's idea would have remained little more than a philosophic dream. Ezra Pound's observation in this context is very apposite.[ii]

It was the discovery of an enigmatic photograph of a woman, casually seated on a slatted park bench, by an unknown photographer, that first kindled my interest in the Woodburytype *(plate F-1)*. My first, albeit mistaken assumption, that it was a gold-toned albumen print that had survived in pristine condition (with no suggestion whatsoever of fading), was incorrect. My initial examination led me to presume that it was a carbon print. However, the precision of the image, in terms of the depth, definition and sharpness of the pigmented gelatin relief, led me to conclude that it was, indeed a Woodburytype; a specimen that appears to encapsulate all the essential identifying characteristics that the process embodies, *in extremis*.

The second event was the discovery, in the Archives of the Royal Photographic Society by Pamela Roberts, of Thomas Bolas' Photomechanical Collection of Walter Woodbury manuscript papers, photomechanical prints and test matrixes, together with several fragile gelatin relief specimens, including one of Rejlander's 'Head of John the Baptist' *(plate 44)*. Without Bolas' foresight and generosity, our knowledge and understanding of the extent of Woodbury's inventiveness and genius may never have been known.

A study of both the British Library and David A. Hanson catalogues reveals the breadth and extent of the range of publications for which Woodburytype was employed: for the reproduction and replication of museum objects, photomicrographs, astronomical records, medical illustrations, art reproductions, promotional and sales material for scientific, industrial, and government reports, agricultural equipment, sheet music, portrait and topographical subjects, museum and commercial art objects, religious propaganda, technical manuals and private press publications. It was used not only for high quality print, but also for high-volume ephemera such as weekly or monthly newspapers.

A number of factors led to its demise: the fact that it was never successfully adapted to print with letterpress; it could not adequately reproduce photographic images with large areas of white; nor could it be used to form images much larger than 10 by 8 inches. However, for a period of fifteen years the Woodburytype process set the standard only occasionally matched by the Talbot-Klic process towards the end of the ninteenth century. As late as 1910, Waterlow's advertisements in trade journals indicated that they were still operating a small facility at Finsbury, providing photographic illustrations for specialised publications, for the British Museum and the Society of Antiquaries, with the requirement for fidelity, precision and permanence being the overriding consideration.

For more than one hundred years, apart from the occasional footnote buried within specialised publications, Walter Bentley Woodbury's contribution to the reprographic arts has never been fully recognised or studied—until the advent of this present work. Barret Oliver's journey has been long and arduous, information relating to the origin and evolution of the Woodburytype being spread far and wide. Casting his net worldwide from his base in Los Angeles, Barret's studies have drawn upon sources from Australia to Austria and Germany to Great Britain. The author has added immeasurably to our appreciation and understanding of one of photography and the reprographic art's great pioneers and until now recondite heroes—complementing and extending the scope of Wachlin's invaluable work, *Woodbury & Page: Photographers Java* published in 1994.

—Michael Gray, Bath, 2006

---

[i] Talbot's Photoglyphic-engraving, Poitevin's Photo-lithography, Photo-collotype and Pretch's Photo-galvanographic processes.

[ii] *Quartets* by Ezra Pound, quoted by Wylie Sypher in *Rococo to Cubism in Art and Literature, Transformations in Style, in Art and Literature from the 18th to the 20th Century*, New York, 1963, 4, 4, 316.

AUTHOR'S PREFACE

Within the past decade or so, the photographic community has experienced a renaissance in nineteenth century processes. The once-forgotten platinum print, albumen print and collodion negative are again currency for serious photographers. At the same time, photography has undergone a revolution with digital technology. Traditional silver-based photographic materials are disappearing in place of electronic media. This has extended to the realm of publishing where plates are made directly from digital files.

These two phenomena are at opposite ends of the spectrum, yet seem to live happily side by side, even feeding off each other. The fruits of these two impulses are nowhere more apparent than in many of the fine photographic books that have been published recently using digital printing technologies to reproduce prints made with nineteenth century processes. Once again the fine photographic reproduction is highly valued and being distributed on a scale never before seen.

It seems as though we have suffered the majority of the past century with inferior reproduction methods and just now are starting to overcome this problem. Technology had reached a satisfactory level and speed of production had won out over superior quality. There have, of course, been many fine and some quite beautiful books published during this time, but not in the vast quantity and diversity that we now see.

One hundred and forty years ago a new method for reproduction was developed: the Woodburytype. It was the first practical method for mechanically reproducing photographs and for two decades was the most widely used. The publishing industry embraced it because for the first time they could include actual likenesses of people or scenes without relying on the traditional method of manually copying onto engraving plates. The photographic community loved it for its high quality and inherent beauty and because they could finally have their work widely distributed. The photo-historian J. S. Mertle called the Woodburytype the most beautiful photographic reproduction process ever invented. [i]

The Woodburytype was the first practical photomechanical process, producing the finest, most faithful reproductions, and has still never been surpassed for quality. It stands as the ultimate in photographic achievement: unparalleled tonal rendition, beautiful delicacy in shadow and highlight areas, supple surface texture, permanent beyond all other processes and accessible to a wide range of people as published reproductions. In some ways they are better than silver prints. Even though the process is laborious by today's standards, the quality achieved from the Woodburytype far outweighs its effort.

◆

Over the course of the past few years I have been engaged in intensive research on the history and technology of the Woodburytype. It was fascinating to discover that this most beautiful process had all but been lost and despite the renewed interest in nineteenth century processes, no one had managed to successfully revive it.

When I first began my research, I was only interested in reproducing the process, but I could find no information in modern literature beyond a brief outline—no accounts of its workings, no details about its mechanics and no thorough history. I had to go back to primary sources to find any practical information. Through period journals, books and original unpublished writings, I amassed an enormous amount of information on the history of the process.

During that time two important things happened that shaped my understanding of the Woodburytype. The first had to do with process and the second with writing about it.

Being a practicing wet-plate photographer, I know something about the difficulties associated with nineteenth century manual photographic process. But nothing could have prepared me for the continuous problems I encountered working the Woodburytype. Having twenty-five years of technical articles from period journals and Woodbury's own notes as my guide, I still had failures at every step of the process. I must have made more than a hundred printing plates before I hit upon the right formulas and combination of techniques. Unlike modern-day collodion, there was no one practicing the Woodburytype to whom I could turn for advice.

It still amazes me how Walter Woodbury was able to work out the process so perfectly with imperfect chemicals and materials and no modern conveniences, not to mention the frequently cloudy weather of England. Even after Woodbury licensed the process to commercial printing firms, they continued to experience problems and had to refine their techniques. That so many of them persisted is testament to the beauty of the process and people's faith in it as the single best method for reproduction.

Gaining such a complete understanding of the process, I was struck by the fact that Woodburytype prints are not at all like traditional prints (woodcuts or etchings), nor are they like silver prints (photographs). I knew this theoretically, but not in practice. The process is truly unique—its key principles are not shared by any process before or since. There is no simple way to describe the virtues of the Woodburytype and what exactly makes it unique—especially to someone with no photographic knowledge.

As a photographer I have always relied on visual images as my primary means of communicating ideas. But I realized I would have to start looking at the prints the way a historian would, and I would have to become versed in the language about communication of ideas, both through words and through images.

I started exploring *syntax*—the visual structure of prints—as described by William M. Ivins. But I still struggled with how the Woodburytype fit in to this schema. Most other writers on the subject mention the Woodburytype only in passing and assign it little historical importance. Of course I have a bias toward the process, but it really seemed to me there was something missing from all the discussions about syntax. Then I realized what it was and vast possibilities opened in my thinking; I discovered that no one had mentioned the reason for the existence of syntax. The reason for syntax is the prime factor in why prints look the way they do. Because the reason derives directly from the workings of the process, it is quite possible that only hands on experience could yield this discovery.

Most people steeped in the history of photography have seen Woodburytypes, but very few know anything about their history, and almost no one knows how to make them. What happened to this first reproduction method? Why was it used for only a very short time? And if it indeed was the most beautiful method ever invented, why did it disappear?

I have written this treatise in an effort to fill a perceived vacuum in existing literature. It is not a comprehensive biography of Walter Woodbury, nor is it a technical how-to manual. I do give a rough guide to these important areas, but instead I have focused on the history of the Woodburytype and why it was special among hundreds of print techniques. It is a general work that aims to provide a foundation on which the Woodburytype process can be studied further.

Because this book uses modern offset printing technology, it is impossible to duplicate all the subtleties of the Woodburytype process. The very nature of the prints is being denied the reader by using a line screen and dots to reproduce the illustrations. The body of this book explains why. To fully appreciate the beauty of these prints, anyone interested in the Woodburytype is urged to look at an original. They are not hard to find; every major collection of art, photographs or books, including university libraries, should contain examples.

I hope that my findings will fill a certain need for collectors, curators, conservators and practicing photographers alike and that they will spark further interest in these truly beautiful art objects and the history of the medium.

As Edward Wilson wrote upon Woodbury's death, "He was only fifty-one years old, and it will be fifty more before our loss can be repaired." I hope that in some part this book helps to repair that loss.

—Barret Oliver, Los Angeles, January 2006

---

[i] J. S. Mertle, "The Woodburytype—The Most Beautiful Photographic Reproduction Process Ever Invented," Image 6 (1957): 165-170.

ACKNOWLEDGEMENTS

I have relied on the support of many people in writing this book—not just in the routine activities of research, but for advice, assistance and moral support. I would like to thank the following people for their generosity, without whom this book would have been an even bigger struggle than it was.

My first thanks must go to Stephen Berkman, friend and photographic partner, who gave me the original inspiration to work on the Woodburytype during one of our all night printing sessions.

My research efforts were generously supported by the following people who gave freely of their time and knowledge. I would like to thank Teresa Mesquit who gave me advice with technical matters; the staff of the J. Paul Getty Museum Department of Photographs, for access to their original prints; particularly Michael Hargraves for his help in locating and identifying obscure material; the staff at the Getty Research Institute Library and Special Collections, for access to their material and research assistance; Dusan Stulik, Getty Conservation Institute, for his assistance with scientific analysis of Woodburytype prints; the Staff at the National Museum of Photography, Film and Television, Bradford, England, particularly Brian Liddy, for his invaluable assistance with the Woodbury archives and Paul Thompson for his excellent work photographing Woodbury's original material; Martin Barnes, Anne-Marie Eze and the staff at the Victoria & Albert Museum for access to their collection; Larry Schaaf for research advice; Steven White for sharing his knowledge about John Thomson and Woodbury; Craig Woods for his help with computer matters; Chris Rowlin and Natasha Mulder at the Science and Society Picture Library, England for their assistance with reproductions.

My special thanks to Bill Dauber for his thorough help editing the manuscript; Sally Tomlinson for her wonderful hospitality in Bradford; Walter Murch (the younger) for his hospitality in London. My extra special thanks to Michael Gray for writing the excellent foreword for this book and for the most warm hospitality he and Barbara Gray gave me with their insight on Woodbury and his associates; Luanne Woodbury, the grandniece of Walter B. and custodian of the family history, for her time, friendly conversation and use of family material in her possession. My particular thanks to Carl Mautz for his patience, time and faith in my little book. My deepest gratitude to Julian Cox, for his continued and invaluable support both personally and professionally, without whom this book would not have been at all possible. And to my family: Kyle, Kathy, Kent, Anita and Gabby.

—Barret Oliver

TIMELINE

Timeline of Significant Dates Related to the Woodburytype Process:

Ca. 1490—Leonardo makes first known nature print

1798—Vauquelin discovers that bichromate salts are light sensitive

1802—Wedgwood and Davy make first attempts to print images with silver nitrate coated on paper

1827—Nicephore Niepce makes the first photo-mechanical print—a reproduction of an etching

1835—Auer, working at the government printing office of Vienna, makes nature prints using lead plates

1839—Daguerre introduces the first practical photographic process; Talbot introduces the first positive-negative process; Ponton uses the bichromate salts to make prints

1852—Talbot discovers the use of gelatin with bichromate and starts working on developments that lead to the photo-gravure

1852—Woodbury travels to Australia, then Java, and teams up with James Page to start Woodbury & Page

1855—Pretch starts working on developments that lead to the collotype and carbon processes

1859—Woodbury returns briefly to England, sells photographs to Negretti & Zambra, returns to Java

1860—Wallis patents his Auto-Typography process using lead printing plates produced from a relief drawing

1863—Wallis writes "The New Art of Auto-Typography" and visits the South London Photographic Society; Woodbury is married and returns to England

1864—Swan perfects the carbon process; around April, presents mechanical carbon prints to Photographic Society

1864—September 23, Woodbury receives his first patent

1864—October 14, Announcement of Swan's process in *Photographic News*

1865—March 17, Woodbury announces his photo-relief printing process in *British Journal of Photography*; July 6, Swan receives a patent for his Photo-Mezzotint process; August 18, Woodbury uses lead plates for the first time, making the process truly practical

1866—The first published Woodburytype is issued in *Photographic News*, *Mountain Dew Girl: Kilarney*, by H.P. Robinson; Goupil buys patent rights for France; the Woodbury Permanent Photographic Printing Company buys the rights to both Woodbury's process and Swan's Photo-Mezzotint

1868—Albert perfects the collotype

1870—Carbutt buys patent rights for America, moves to Philadelphia and starts new shop; Franco-Prussian war destroys Goupil's shop; *Photographic Art Journal* starts publication; Woodbury, in conjunction with Goupil, develops the first practical photo-gravure process

1872—Woodbury edits *Treasure Spots of the World*; Goupil finishes rebuilding and starts mass production of Woodburytypes

1877—Thomson & Smith release *Street Life in London*

1880—Woodbury perfects the Stannotype, which does not rely on the hydraulic press, making the process practical for amateurs

1885—September 4 or 5, Woodbury dies at Margate of an overdose of laudanum

Ca. 1890—The halftone process is perfected, superceding all existing photo-mechanical process for mass production

1928—Eder sells last Woodburytype press for scrap metal

A HISTORY OF THE WOODBURYTYPE

The First Successful

PHOTOMECHANICAL PRINTING PROCESS

and Walter Bentley Woodbury

1

Walter Bentley Woodbury —The Early Years

Some men labour, others enter into their labour.
—Walter Woodbury's epitaph

Walter Woodbury was a prolific experimenter, devising a multitude of photographic apparatus and techniques, but an inept businessman, losing the rights to his patents and all his money and businesses. He was an adventurer, spending years in exotic locations, but suffered ill health throughout his life. He was a fast friend to many in the photographic community around the world, but a persistent enemy to anyone who he thought had cheated him. He was famous among the photographic community in his own lifetime, but nearly forgotten after his death.

Most of all, Woodbury was an inventive genius who understood not only the mechanics of photography and its true potential, but the direction it would take in the future. In the early part of his career he lived the typical life of a Victorian frontier photographer. As his photographic skills matured, his contributions changed the way photography was practiced. He was the last in the line of inventors who worked in photography as a manual craft and the first to turn out mass production images. He occupies a unique place in photographic history, as do the processes he developed—the connecting link between the manual and the industrial. He died at age fifty-one in 1885, and one wonders, if he had lived, where his keen mind and adventurous spirit might have taken him as the photographic world moved into the twentieth century.

Walter Bentley Woodbury (frontispiece) was born on June 22, 1834 in Manchester, England. His father was John Taylor Woodbury, the son of a warehouseman, and his mother was Ellen Bentley, daughter of Walter Bentley, for whom Woodbury was named. Walter Bentley was a well-known amateur naturalist famous for his reconstruction of an elephant skeleton that stood in the South Kensington Museum for a number of years.[1]

John Woodbury died at the age of thirty-six, when Walter was only twelve. Two years later, Ellen married James Lea, seven years her junior, who died at the age of thirty, only two years after the marriage.[2] Around this time Walter was sent to live with his uncle, where he received a meager private education.[3]

He had an early love for the camera and, in his youth, entertained friends with the novelty whenever he could. In a later memoir he wrote:

> *My greatest pride, at the age of twelve, was to take my boyish companions to the garret, at the top of the house, where I had, by means of an old magnifying mirror, from which I removed the mercury backing, and a sheet of looking-glass, been able to throw a two-foot image onto a table beneath. . . . The idea that such a picture could be fixed was too wild to be even dreamt of in my young philosophy.*[4]

One person recalled he "appears to have never been without a camera since he was old enough to carry one."[5]

His first job was as an apprentice in the Manchester patent office, where he trained to become a civil engineer. But the direction of young Woodbury's life changed in February 1851, when E. Hargraves, a California miner, discovered gold twenty miles north of Bathurst, in the Macquarie Plains of Australia. This discovery, only a few short years after the one in California, caused a major sensation and created an enormous shift in the Australian population and production base. Crews abandoned perfectly good copper mines to make the transition to gold sites, masses of people migrated to cities from rural areas, and foreigners arrived by as many as 2,000 per week. The province of Victoria increased the value of its imports and exports ten times between 1851 and 1853.

Many people left England during this time to find their fortunes in what was still a colonial outpost. Bitten by gold-fever, Woodbury soon traveled to Australia. He sailed from Liverpool aboard the *Serampore* on July 6, 1852, arriving in Melbourne after a three-month journey.[6]

He quickly discovered mining did not produce the kind of profits he imagined, and contrary to almost all the published accounts of Woodbury's time in Australia, he never actually engaged in mining as a living. He wrote prolifically to his mother during his stay. In the first letter he described his arrival. "I did not go to the diggings as I intended because there was very bad news from there, not one in a hundred being able to make as much as would pay for their food."[7]

Woodbury came from a solidly middle-class background, but soon learned to fend for himself by working manual labor jobs. Again, he wrote to his mother:

> *Under these circumstances I preferred getting a situation in Melbourne which I found rather a difficult thing to do so I was obliged to pick up with a man to drive a team of horses to the diggings with £3 per week and my food. As you may guess I found this rather hard work. I had to sleep under a cart at night on the hard ground with a sack of oats for a pillow and to walk about 20 to 30 miles a day. . . . In the short time I have been here I have roughed it more than I ever did in my life before.*[8]

Finding work in this climate proved to be a great endeavor. His life in Australia was one of constant change, going from job to job, as work was available. He even spent a week as a cook, before finding employment as a surveyor's laborer. He wrote his mother that December:

> *I then tried to get something else but found it was rather a difficult job. . . . I was above a month without anything to do. . . . I would not advise anyone that has a good situation to give it up to come out here, thinking to make their fortune in a few days, as they will soon find out their mistake when they get here. They must either work hard or starve. I do not in the least regret it myself, as when I get home again, if ever I do, I shall be able to turn my hand to anything and not be afraid of hard work.*[9]

The job with the surveyor, Mr. Dawson, proved to be a steady position and Woodbury managed to save a little money. By January 1853 he wrote his mother that he had again picked up a camera, "I also amuse myself with a Camera Obscura that I have bought. I can at least manage to take a good likeness, now I never have a failure.

I am at present instructing my master in the art of Photography."[10] It is unknown whether Woodbury was using the older salted-paper process or the new collodion process, introduced in 1851; he eventually became known for his skill as a collodionist.[11] A later recollection from Woodbury slightly embellishes the event:

> *I found my little means almost exhausted, being reduced to about £6. . . . But just at this time, of all extraordinary things, what should I see in an old store-shop but a camera and lens for sale. My old love returned at the sight of it—it had evidently never left me—and I, recklessly, never thinking of the question, "What will I do with it?" went in and spent two-thirds of my last remaining worldly wealth in its purchase.*[12]

Woodbury was very fond of his employer, whom he mentioned often in his letters home. The wages and camaraderie of the job enticed Woodbury to stay with the survey party for about a year before he found a position in January 1854, as a draftsman at the Commission of Sewers and Water Supply for the City of Melbourne.[13]

During his time at the water works, Woodbury managed to have enough free time and extra money to invest in his photographic hobby. He made portraits for his friends and "views" of the city and surrounding country. He became quite skilled as a photographic artist and wrote his mother in December 1854, "I got the second prize medal for photographic specimens in the exhibition."[14]

In the same letter, Woodbury told his mother that he again was without a job, the load at the water works declined and they had to let him go. But this time he wasn't looking for another position. Instead he looked in another direction, "the only work to be done in Melbourne is either stone breaking or shepherding neither of which I feel inclined to take so I am going to turn photographic artist."[15] Later Woodbury recalled, "some one, wiser than myself, one day suggested that I was spending all my little earnings in buying chemicals and glass without any remuneration for myself, and that I ought, at least, not be the loser."[16]

In Melbourne he had met another young Englishman named Spencer who had photographic inclinations and the two were to open a traveling studio together, going out to the mine camps to take likenesses for the miners. "We are not going to start in

Melbourne as there are already about 20 [photographers] in it, but are going to buy a horse and dray and tent and make a tour through the country."[17] This arrangement did not last more than a few months, as the partners had as much trouble getting portrait business at the mines as the miners had finding gold. Woodbury and Spencer separated, with Woodbury returning to Melbourne and setting up a little shop of his own and Spencer staying on at the camps with only mild success.

Shortly after, Woodbury received an offer from a local Daguerreotypist, Mr. Batchelder, to join his studio. Batchelder was eager to start working in the new collodion-on-glass process, a faster and more economical photographic medium than Daguerreotypes, which were starting to fall out of fashion. Woodbury wrote to his mother, "Mr. Batchelder. . .had come to the conclusion that I was the best glass artist in Melbourne," and that "[my photographs] are acknowledged by all who see them to be superior to anything in England or America."[18]

This job gave him the most stability and income of all his endeavors to that point. He was now making £4 a week, had rented a little house and was running his own portrait business in his off-hours in the morning.[19] And at that time, contrary to his earlier admonitions, urged his mother to send his younger brother Henry James (or Harry as Woodbury sometimes called him) to join him in Australia. "What a valuable assistance Henry would be to me here. . . . I might as well be paying him as a stranger."[20]

He soon left Batchelder's establishment, putting all his energies into his own shop, with a new assistant, Harry Davis. Woodbury wrote his mother that he had built the studio *(plates 1 and 2)* entirely by himself.[21] The studio was too far from the center of town, however, which restricted business to the immediate neighborhood, and he resolved to move.[22] Woodbury and Davis left Melbourne entirely, settling for a time in Woolshed Creek where they set up another small portrait studio.

It was here that he met James Page, "who I had taken a great liking to, more so than anyone else I have met in my travels."[23] The two soon joined forces and, because of slow business in Australia, departed for Asia. They stopped in Java (Indonesia) and cancelled plans to travel further when they found a wealthy European population, cheap labor and small photographic community. Their portrait studio, Woodbury &

Page, did a brisk business from the beginning, "every day we have had carriages at the door every ten minutes and have got engagements for the next week."[24]

Most of their work was divided between glass pictures (ambrotypes) and paper prints. "We get very good prices for our portraits—from 20 Rupees (1£-14s) to 120 (10£). Those we charge £10 for are done on paper and afterwards colored by an artist here."[25]

They did not stay in the principal city of Batavia the entire time. They traveled about taking portraits, and when the opportunity arose made views of the country and its native inhabitants *(plates 5-9)*. Woodbury sent photographs to the *Illustrated News* in London, which reproduced one of his still-lifes *(fig. 1)*.

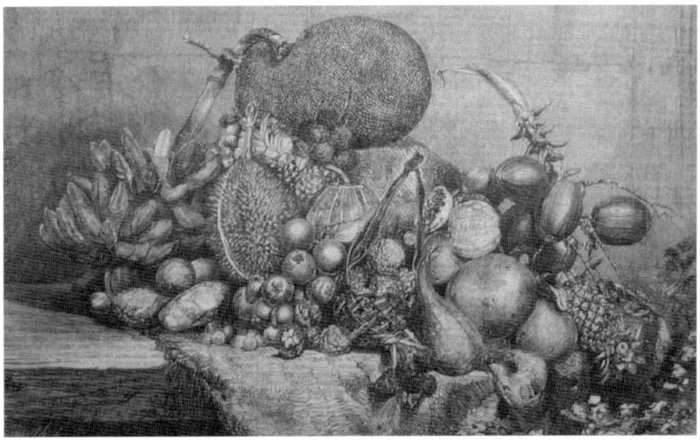

Figure 1. Group of Indian Fruit from the Island of Java, *from a photograph by Woodbury published in* Illustrated London News, *October 15, 1859. Notice the striking similarities to plates 37 and 38. It is quite possible the engraver for this print copied it from a similar still life sent by Woodbury.*

Details of Woodbury's stay in the South Pacific become cloudy at this point. In the last surviving letter to his mother he wrote he planned to travel home as soon as Henry arrived to take over his place in the business. Amid rumors of ill health and a straining partnership with Page, Woodbury said the purpose of the journey was to secure photographic supplies, which were extremely hard to get so far from the centers of manufacture, and sell a stock of images to firms in England and Holland (under whose control Java was at the time). He did indeed return to England sometime before the end of 1859 (but the dates are unclear) and secure supply lines. He also sold many pictures to the firm Negretti and Zambra, who felt the photographs "in point of artistic effect, we believe to be unequalled."[26]

Sometime around January 1860, Woodbury returned to Java and expanded the portrait business. By June he set out on an extended photographic journey away from the studio and wrote a three-part article for the *Photographic News* detailing this excursion, during which he experienced all kinds of difficulties with the climate, the terrain and the photographic equipment.[27]

For a time he broke partnership with Page and operated under the name Atelier Woodbury *(plates 3 and 4)*. But before Woodbury's ultimate return home, the two united again and Woodbury & Page re-opened. Woodbury's other brother Albert also joined him in Java and participated in the studio business.[28]

During Woodbury's residence in the South Pacific, he met Marie Sophia Olmeijer, daughter of a Dutch father and Malay mother. Their courtship was thought unwise by Woodbury's English family and acquaintances because of Marie's mixed blood and the extreme prejudices against inter-racial relationships at the time. Henry begged Walter not to go through with a marriage.[29] But the two were married in January 1863, departing shortly for England, after which Woodbury never returned to the South Pacific.[30]

Shortly after Woodbury's return to England, James Page died, leaving the business in the hands of Woodbury's brothers. It continued to operate, under changing names and ownership, until 1908, twenty-three years after Woodbury's death.[31]

From the time of Woodbury's return to England until 1864 he worked sporadically on the developments that would later become the Woodburytype. His experiments started with two observations. On his trip to England in 1859 he saw a window transparency hanging in a friend's home and was fascinated by the different opacities it possessed.[32] Then, in Java while making prints by the carbon process, Woodbury noticed the distinct relief in these pictures. He put the two ideas together, thinking a carbon print used as a mechanical plate should be able to produce different opacities like the window transparency.[33] He began his search for suitable materials to make what some had called mechanical carbon prints.

But it seems this was not yet his overriding passion, for when Woodbury announced his process, he was still trying to make his name as a photographer, not an inventor. In February 1864, the same year as his patent, Nigretti & Zambra issued a series of stereoscopic slides of Woodbury's photographs from Java.[34]

The *Photographic News* commented:

> *The subjects are charming, and the negatives evidently very perfect, indeed. Whether for instruction or amusement we cannot well conceive a more valuable series. Some of the scenes are exquisitely beautiful as pictures, foregrounds with gorgeous profusions of tropical foliage, added to fine river scenery and delicate distances, producing an effect which a poet or painter might imagine, but which we should scarcely hope to meet in a realistic transcript of natural scenery.*[35]

There is a deeper symbolism here as well that has to do with quality. Early photographers were obsessed with quality. Most of the inventions or variations in technique, from the waxed-paper process of Le Gray to the collodion wet-plate and albumen papers, were aimed at increasing the quality of the image. Paper negatives could be used for making multiple prints, but they could never compete with the clarity, brilliance and verisimilitude of the Daguerreotype. After photomechanical technology became a reality, quality steadily decreased in reciprocal to ease. With each advance in speed came a loss of detail and beauty. The Woodburytype is the apex in this curve; the culmination of research that began with Niepce, and the quality has never been surpassed by subsequent developments.

The Woodburytype links these two periods chronologically, but it is also a peculiar blend of technology from the two divergent worlds. It was the first mass production photographic technology, but still required competent craftsmen to run the hand-powered machinery, and the first stages of the process demanded a skilled photographer with a discerning eye to know when the plates had been prepared properly.

A single invention may not have caused all these changes directly, but as the first photomechanical process to be worked on a large scale, the Woodburytype certainly contributed to them.[37] With a relatively short life span of about forty years, it permanently changed the idea of photographic reproduction that had been a dream since the fathers of photography first started experimenting.

◆

On March 17, 1865 Walter Woodbury announced his new process of photomechanical printing *(fig. 2)*.[38] Photographers had searched, since the beginning of the medium, for a practical way to reproduce images in large quantities. Using extant technology, photographs could be printed only in small editions through an arduous process of messy and sometimes dangerous chemical treatments. The traditional printing techniques of engraving and woodcut provided the best example of multiple-printed images, but could not translate the subtle details and rich tones of a photograph into a printable surface for mass production.

WOODBURY'S NEW PRINTING PROCESS.

I BEG to submit to your notice an entirely new method of photographic press printing, which, although in the specimens sent is not so perfect as might be wished, still promises the greatest success. You will see that it is entirely different from any press photographs that have hitherto appeared, in the fact of its appearance being exactly the same as an ordinary silver or carbon print, and not possessing any grain or stipple. This, I think, is likely to prove a more simple and efficient method than any hitherto brought before the public. I will, therefore, give you a brief outline of the method of proceeding by which the accompanying pictures have been produced.

A copper mould is obtained by exposing a sheet of bichromated gelatine behind a negative, and exposing to parallel rays of light; then washing away the undissolved portions and taking an electrotype cast in copper. This mould is then backed with gutta-percha and placed in a suitable press, having a sheet of plate glass, with strong springs, let into the back. Having greased the mould, a transparent mixture of gelatine, water, and suitable colour is warmed, and a small quantity poured on the centre of the mould. A sheet of paper is then laid on it, and the lid of the press bolted down. This, by exercising a steady pressure, squeezes all the superfluous colour out at the side; and on opening the press, the picture is taken off the mould, and in that state is in relief, which, however, dries down to be very slightly so.

The operation is remarkably quick. By having three presses, and, after setting one, then going on with the others, the picture in the first will have set; and by continuing this, as many as 200 may be easily turned out in an hour. The expense also is very trifling, as all the extra gelatine and colour squeezed out may be used over again. The principal things necessary to success are—the most exact surface of the mould, which, if bent in the slightest degree, would produce an uneven picture; also, the addition of the colour in the exact proportion to produce the necessary depth in the picture.

I hope to be able to send you better specimens and complete details of the process in a few days. It is a curious fact that having, in the earlier stage of my experiments, shown a rough specimen by this process to a friend of Mr. Swan's, he acquainted me with the fact that that gentleman had also had a similar idea to myself; but, as he learned that I had patented the process, he did not proceed with it. Still, as the idea of the thing occurred to him, independently of any information from me, credit is equally due to that gentleman.

WALTER WOODBURY.

[We look forward with much interest to the working details of Mr. Woodbury's extremely promising process. The specimens he has forwarded to us cannot be distinguished from ordinary silver or carbon prints taken direct from the negative, except by a slight indistinctness of outline, which may, however, not be a defect inseparable from the process. There is no perceptible grain whatever, and the deposit seems to be as fine as reduced silver.—EDS.]

Figure 2. The first published article describing the Woodburytype published in British Journal of Photography, *March 17, 1865.*

In the first few decades of discovery, inventors developed two distinct ways of producing a photographic image: photographic prints and photomechanical prints. These categories are functional only, but a distinction between them helps in understanding the different avenues of inquiry.

A photographic print (or photograph) is one produced by the direct action of light upon a sensitive surface which produces a likeness of the image. It may be made by enlargement or contact and usually includes the use of a camera negative as the subject to be reproduced. The light-sensitive material is usually a metallic salt, which upon exposure, becomes reduced to a base metal and appears dark. Each individual reproduction requires its own exposure to a light source, sometimes being put through a series of chemical baths before the image is revealed.

A photomechanical print makes use of ink (or similar material) upon a surface (commonly a block or plate) affected with a tactile pattern to alter the intensity of ink in proportion to the pattern of the image. It is also usually made from a camera negative. It always makes use of light to produce the pattern on the printing surface, but never in the transfer of ink to the final reproduction. Any of the printmaking methods may be used (i.e. intaglio, relief, litho, etc.) but all produce a finished image in ink.[39]

The first documented experiments with photography were done by Thomas Wedgwood—the son of Josiah Wedgwood the porcelain maker—and (Sir) Humphry Davy. They titled their groundbreaking paper of 1802, "*An Account of a Method of Copying Paintings Upon Glass, and of Making Profiles, by the Agency of Light Upon Nitrate of Silver.*" The primitive camera obscuras of the time were not capable of capturing images directly, so Wedgwood and Davy concentrated on copying existing pictures. They used a method called superposition, or contact printing, in which the painting or print is laid directly in contact with a surface coated with silver nitrate and exposed to the sun. As the sun struck the sensitized surface in proportion to the image, darkening the silver nitrate, they produced a negative or reverse of the original. Wedgwood wrote:

> *The consideration of these facts enables us readily to understand the method by which the outlines and shades of painting on glass may be*

copied, or profiles of figures procured, by the agency of light. . . . For copying paintings on glass, the solution should be applied on leather.[40]

Unfortunately, they were unable to fix the images, which continued to darken as more light struck the surface. As unsuccessful as Wedgwood and Davy were, they laid the groundwork for all photographic technology that followed and formed the model for future image reproduction.

A little more than a decade later, Nicephore Niepce started the experiments that would become his *heliographic* process and, after his death, formed the basis of the Daguerreotype. Because camera obscura technology was still inadequate, he succeeded in reproducing an existing image, rather than producing an image directly from the camera, which he did later. Like Senefelder, the inventor of the lithograph, Niepce was searching for a way to reproduce a unique work of art without the intervening hand of manual copying onto a printing plate. Niepce used etching as his model and starting point. He coated a metal plate with a light-sensitive etching ground that hardened when exposed to the sun and then dissolved the unexposed areas, producing a negative. The plate was then etched with acid to make pockets that would hold printing ink. With the aid of a professional printmaker, he succeeded in producing the world's oldest surviving photographic image and the first photomechanical print *(plates 10 and 11)*.[41] J. Towler, author of the classic photography manual *Silver Sunbeam* later wrote:

It is a curious fact that experiments in photographic engraving gave rise to photography itself. The idea, the most prominent in the mind of Nicephore Niepce, when he commenced his indefatigable researches in 1813, was not only to fix the images obtained by the camera obscura on a plate of metal, but to convert this plate into an engraving from which to receive prints from the press.[42]

In his "Notice on Heliography," Niepce specifically writes about copying engravings:

There is doubtless nothing difficult in employing the means of execution which I have just mentioned, yet nevertheless the first attempt might not be attended with success. I think therefore that it would be advisable to begin by copying engravings.[43]

Daguerre had been working on perfecting the camera obscura and when he partnered with Niepce, the emphasis of their experiments shifted away from photo-etching and toward camera images. When Niepce died, this process might have been lost without his son, who insisted that Daguerre fulfill the obligation to his previous partner. When Daguerre published the official report on the Daguerreotype, he included Niepce's essay "Notice on Heliography."

The beauty of the Daguerreotype captivated the world, but it was not much of an advance for reproduction. Because Daguerreotypes were single positive images that came direct from the camera, they had to be re-photographed if copies were wanted. They were also made on stiff metal plates that could not easily be inserted into books. But this drawback presented a possibility. Many people tried to turn the Daguerreotype plate itself into a printing surface. Hippolyte Fizeau developed what was probably the most successful method. Through a series of arduous chemical treatments, an original Daguerreotype plate was etched and electroplated, and could be used to make a print in much the same manner as a traditional etching. Charles Chandler, author of many books on photomechanical printing, wrote:

These pictures were taken on silvered copper, and it was soon found that electrotypes could be made from them which would show the picture in relief. These reliefs were not, however, sufficiently deep or sharp for practical use with printer's ink.

It was the development of Talbot's invention of the negative that finally revived photo-mechanical work, and made it possible to secure satisfactory results.[44]

William Henry Fox Talbot independently developed a photographic process at the same time Daguerre was working on his metal plates. The Talbot process, however, was based on a different approach; he followed Wedgwood's and Davy's idea of making images on paper. Like his predecessors, Talbot copied existing images.

He wrote:

> *The Shadow-pictures which are formed by exposing paintings on glass to solar light are very pleasing. . . . The pictures thus formed resemble the productions of the artist's pencil more, perhaps, than any of the others.*[45]

Talbot began by making contact prints with botanical subjects, which yielded delicate, beautiful results that opened a new world for disseminating scientific information.

> *The first kind of objects which I attempted to copy by this process were flowers and leaves. . . . These it renders with the utmost truth and fidelity, exhibiting even the venation of the leaves, the minute hairs that clothe the plant, &c.*[46]

Although these images and those he later produced from the camera were negatives, Talbot soon discovered he could copy these negatives by superposition and obtain positive images. Then he found that a negative exposed for a very short time and still not visible could be developed by subsequent chemical treatment. These two developments led, more than any, to our modern film system, and also to more avenues for photomechanical processes.

The problem with Talbot's process was that in making a positive print the light had to pass through the paper of the negative, which made the image fuzzy. Various methods were tried to overcome this problem. The best advance came in 1851 when Frederick Scott Archer developed the collodion-on-glass, or wet-plate, process. This new method was quicker than all previous ones and, when used with albumen coated papers, produced extremely detailed and clear images that were the standard for generations to come. It superceded the Daguerreotype and the paper negative as the method of choice for making in-camera images.[47]

Continuing his experimentation and employing discoveries by other inventors, Talbot found that if he printed a negative through a screen, it fragmented the image into tiny dots. The dots varied in size depending on how much light penetrated the negative. In the highlight areas the dots were very small or non-existent. In the shadows they were larger or merged together completely. He used this together with a light-sensitive etching ground, much like Niepce had done, to produce a plate with tiny pockets that could hold ink. He called this process *photoglyphic engraving.*[48]

Other processes for photomechanical printing were developed around this time such as photo-etching, photo-typography and photo-lithography *(plate 12).* But none of them had advanced sufficiently by the time the Woodburytype process was developed. These early photomechanical processes were a step in the right direction, but had not solved some of the major problems needed to make them practical.[49]

The most obvious drawback was visual. The translation of a photograph into a printing plate caused the original continuous tones of the photograph to be replaced by small pockets that would hold ink. Talbot's photoglyphic process, Fizeau's process and photo-lithography are all based on this premise. This was accompanied by a loss of detail that made the print look more like a drawing than a photograph. It was the same kind of loss of visual brilliance that contributed to the disuse of the paper negative and sentiments changed little as the years went on.

The biggest problem, however, was the lack of durability in the printing plates used in these processes. If the photographer needed only a few prints these worked fine, but for time and aesthetic value, he would be better served with a photographic print. When commerce entered the picture, none of these processes could compete with the ever-increasing demand for speed and quantity brought on by advances in text printing.

◆

By the mid-1800s the industrial revolution was under way with new products being manufactured in large quantities to supply a growing populace. The invention of the printing press and movable type in 1439, with its ability to easily produce multiple copies, has been called the first invention of this era.[50] Rising literacy rates, coinciding with growing ideas about democracy and the importance of the individual, caused the publishing industry to boom. New, faster machines printed everything from simple handbills to richly-illustrated books.

The traditional printing processes of engraving and woodcut provided most of the illustrations of the day.[51] Discerning consumers prized engravings for their extremely fine detail and artistic appearance, while woodcuts were more coarse but could be locked-up with type in the press and printed at the same time. A decision of which to use hinged on aesthetic and economic concerns. A major drawback to these long-established techniques, however, was their dependence on being manually copied from an original in another medium. All these processes were akin in some way to drawing. If a publisher wanted to reproduce an existing image, he hired a draftsman to copy it to a printing plate by hand. Often, a second person did the engraving. This was extremely time consuming and relied on highly skilled artisans. When the plates hit the press several generations later, the copies were not always faithful.[52]

Photography revealed another limitation in the traditional methods. Printing and photography, two graphic media devised on entirely different technologies, were incompatible. None of the printmaking processes could reproduce photographs. As a new wave of images came from the camera, publishers who eagerly wanted to reproduce them in books had no means to do so. Only two options were available; either they inserted photographic prints or resorted to the somewhat backward practice of manually copying them onto printing plates, in much the same manner as for paintings. Photography was supposed to replace manual drawing, if only in its fidelity, so this practice was distasteful to many.

Although closer in verisimilitude, the process of making photographs didn't hold any great advantage over traditional prints for illustrative reproduction. Photographs were just as expensive because of the highly skilled printers and the costly silver needed to make them. But it wasn't just the expense; it was also time consuming and unpredictable. Photography was still a very messy, manual affair. Some commercially produced goods were available, but for the most part, photographers still made their own materials and each practitioner had their own variations in process.

Photographers ran their plates and printing papers through a series of chemical baths to produce a light sensitive surface. The chemical reaction that occurred each time a plate or paper was put in the bath changed the make-up of that bath. The silver (or sensitizing) bath was particularly susceptible to these changes. Also, because electricity was not a common utility and the early photographic materials were only sensitive in the ultra-violet end of the light spectrum, photographic printers used the sun to expose prints. Slight variations in intensity of light caused longer or shorter exposure times. This made each print slightly different than the previous one, necessitating a variance in the printer's methods in order to compensate. With a run of 100 or more, the chemistry would have changed so much that distinct differences could be noticed between the first and last prints.

Another problem faced when using photographic prints was their lack of permanence. Fading, darkening and spotting all occurred because photographic technology had not advanced sufficiently *(plates 13 and 14)*.[53] Woodbury wrote:

> *Had the process of photography as at present used by us been practiced by the Egyptians . . . we should now remain in entire ignorance of the fact, owing to our knowledge that even our most permanent processes would not have been able to stand the test of so many years.*[54]

The photographer and chemist Henri Victor Regnault, chairman of the Paris Photographic Society, said:

> *Of all the elements with which chemistry has acquainted us, carbon is the most permanent, and the one which withstands most of all chemical reagents at the temperature of our atmosphere. The present condition of old manuscripts indicates that carbon in the form of lampblack on paper remains unchanged for hundred of years. If we could therefore make it possible to reproduce photographic images in carbon, we should have a basis for their permanency, as we now have in our books, and that is as much as we may hope for and wish.*[55]

Gum bichromate, bromoil and the *carbon process* all emerged from the search for a pigment photograph. But again these produced only one-of-a-kind prints and were subject to the same problems of manipulation as other photographic processes.

The Woodburytype also used an organic pigment less liable to fading.[56] It was, in contrast to the other pigment processes, designed for mass production; it was mechanical, comparatively swift, less expensive and, even though in monochrome, capable of faithful reproduction without manual intervention. At the same time it was an advance in quality. The Woodburytype has an inherent visual beauty that attracted the buying public as well as the publishers. Its delicate tonal range, ability to capture detail, complete lack of grain, soft surface sheen and pleasing tint make it an art object in its own right. To the untrained eye, Woodburytype prints look exactly like true photographic prints. George W. Simpson, editor of the *Photographic News*, wrote:

> *By Mr. Woodbury's process absolute facsimiles of silver prints are produced, so perfectly that experts may be easily deceived by them.*[57]

And J. Waterhouse, head of the government printing office in India, wrote:

> *Mr. Walter Woodbury succeeded in solving the problem of producing prints without grain . . . and . . . has been able to produce absolutely permanent prints with such perfect photographic gradation, combined with the most exquisite transparent delicacy and richness of tone, that none but the initiated would know that they were not ordinary silver-prints.*[58]

With the development of the Woodburytype, Woodbury achieved what no one had done before him; it was the first convergence of the two graphic technologies, the press and the photograph.[59] Photographs could finally be printed mechanically for inclusion in books.

existing artwork, including photographs. Because the tonality changes very little, all the subtleties of a painting, for example, can be captured and translated to the print. The delicacy of the tones makes it ideal for portraits and the ability to separate detail makes it perfect for landscape and architectural work.

The physical surface characteristics of the Woodburytype closely resemble the albumen prints in favor among photographers in the late nineteenth century.[61] The glossy appearance of these prints makes even the most minute detail clearly distinguishable. Comparatively, a matte-surface print, like an etching, lithograph, photogravure or collotype, softened details, sometimes excessively. While this was occasionally desirable, particularly in portraits *(plates 15a, 15b and 16)*, this effect could be produced in the camera, rather than being imposed by the print medium.

Because the printing ink is made specifically for the process, it may be customized to suit the desires of the printer, client or original artist. Even though the image is monochrome, Woodburytypes were rarely printed in black. Pigments like red, brown and violet were commonly added to approximate the tint of albumen prints. Or the ink could be made to match the color of a drawing or other graphic work.

The most distinguishing characteristic of the Woodburytype, and the one that sets it apart from all other mechanical printing methods, is its complete lack of grain *(plates 17 and 20a)*. Because the printing plate, or matrix, is made from a photographic relief, there is no translation of the image tones, giving the final print a seamlessly modulated appearance.

The grain structure typical of a photomechanical reproduction is a kind of pictorial syntax.[62] This idea of a structure in printmaking was outlined by William M. Ivins in his book *Prints and Visual Communication*. Syntax is the concept postulated by Ivins to describe the systems of translation needed to produce the illusion of gradation in a printed image. Traditional printers' ink is opaque and incapable of gradation. That is the reason for syntax. For the black ink to appear as modulations in tone, the image must be deconstructed into distinct spaces of black or white. The proximity of the solid black dots or lines dictates the shade; closer lines appear as a darker shade, while lines spaced further apart appear as a lighter gray *(plates 18, 19, 20b and 20c)*.[63]

The Woodburytype, however, is printed not with an opaque ink, but a semi-transparent one. The transparency of the ink allows the white paper base to show through the pigment more or less depending on how dense the ink is at that point. The combination of the pigment and the white base produces a true gray value in the middle values, or half-tones as they were called, without having to deconstruct the image. W. K. Burton wrote:

> *I have placed the Woodburytype process before all other photo-mechanical processes, because, as I say, it is the only process which can be quite strictly termed a half-tone process. All the other processes which give a half-tone are printed in opaque ink, and the half-tone is got, as described above, in the case of engravings, some means being used to break up the surface of the block into lines, dots, stipple, or grain of some kind.*[64]

Although not known by the term syntax at the time, nineteenth century artists and printers wanted to eliminate or at least reduce the evidence of this manual intervention in the printing plate. Some of the world's greatest artists worked directly on printing surfaces in order to circumvent the interpretation of their work.

True photographs as prints completely eliminated the manual syntax associated with traditional printmaking.[65] However, this did not solve the problem of reproduction for illustration. Even if the early photomechanical processes, like Talbot's photoglyphic printing or Fizeau's etched Daguerreotypes, could be made to withstand large print runs and produce an advance in syntax refinement, they were unable to eliminate it all together because they relied on image deconstruction like the traditional printmaking methods. This was the obstacle to be overcome in producing a workable technology to print photographs on a press. The *Photographic News* wrote:

> *In photographic printing, gradation is obtained on a different principle: lines or dots, hatching and stippling, are unnecessary. The gradation is obtained by degrees of depth in a continuous tint, like "washing," in water-colour drawing. It is the nature of this gradation, and its dissimilarity from that employed in ordinary printing, which has been the great difficulty in obtaining successful press photography.*[66]

Because the Woodburytype uses a special semi-transparent ink, which is capable of gradation, like the washes mentioned above, and a variable depth plate, it can reproduce every transition in tone without deconstructing the image into a pattern. There is no pattern and no print syntax. Again, the *Photographic News* wrote:

> *It is obtained here by different thicknesses of a coloured translucent material on a white ground. The deepest blacks consist of a sufficient thickness of this material to become practically opaque . . . the next depth of tint is obtained by a slighter thickness . . . the other tints by varying thicknesses, the thinner the layer of the coloured translucent medium, the more delicate the tint of grey, as the white ground becomes more and more apparent, gradually merging into the pure white of the ground.*[67]

Mechanical reproduction methods developed after the Woodburytype returned to the traditional printers' ink because of its ability to be used with and be printed at the same time as text. These methods, which are essentially industrialized versions of traditional printing techniques, all have their own syntax that contributes to the degradation in quality of the original image.

The Woodburytype is the only mechanical reproduction method to contain no syntax of its own and no print pattern. Because the special ink used is capable of gradation, no system of translation is needed to deconstruct the image into distinct spaces of black or white. Or as John Thomson said, "the semi-transparent ink rises in beautiful gradations through the delicate shades and half-tones."[68]

4

The Woodburytype & The Birth of Visual Democracy

To bring by means of photography, the characteristic expressions of our great men, under observation of all, at whatever distance socially or geographically.
—Preface to *Men of Mark*, 1876

The spread of visual information is the single strongest element that shapes our culture today. The photograph, the most accurate of all means of visual representation, is largely responsible for this. It gives us reliable information we would otherwise never have, it produced the cult of personality on a world-wide scale, and is an almost indistinguishable substitute for memory, a great teaching aid, a tool of propaganda, art medium and proxy for anything we cannot experience in person.

The ascent of photography into the modern age spurred remarkable social change. With it came the ability for people to experience, at least visually, things that previously demanded a firsthand witness. You no longer had to travel to Asia to know what it looked like, you no longer had to meet a person to see their face. Photographs of slaves in the American south, of the slums of London and of primitive cultures brought a new social consciousness, while images from the microscope, telescope and hard to reach areas of the globe provided new material for scientific investigation. Because photographs capture moments in time, the past merged with the present in a new continuum. Instant photographs froze motion into analyzable single events, changes in the environment could be compared side by side, portraits of loved ones stayed with relatives long after death.

Although photography was still a young art, its practitioners were bolstered by the acceptance it had gained with the general public. Recognized artists and leading political figures supported it as the new medium of choice. In the post-

Enlightenment era, photography satisfied the values of discovery and accuracy. With the new middle-classes it also catered to the vanity of people who wanted to display their wealth. It was the perfect combination of technology and art and a favorite pastime of the ambitious and the wealthy alike.

Ironically, an article published in 1865 asked, "With respect to Mr. Woodbury's printing process, the question will naturally rise—Will it affect the everyday practice of professional photographers?"[69] Nothing could be answered more affirmatively. Today most images shot by professional photographers are intended for print. It is impossible to imagine a world today without photographic reproductions. It's also impossible to estimate the number printed each year. The average person in any industrialized nation sees dozens, if not hundreds of mechanically printed photographs every day in magazines, newspapers, books, billboards, cheap circulars and now digital prints. Yet only 140 years ago this feat was a marvel.

Like the carte-de-visite a decade before it, the Woodburytype changed the way people lived with photographs. For Victorian readers accustomed to seeing traditional prints in books, which were re-interpretations of the original, or photographs hung on walls, photographs in books were something novel and special. Not only did the reproductions afford a more accurate study of a subject, but their portability had the added value that it could be shared with others. Liberated from re-interpretation, illustrations now had a higher scientific and social value.

During that time people saw photography as an extension of literature—as a way to distribute information, both aesthetically and as document. With photomechanical printing, and the Woodburytype in particular, a whole new era emerged. It made possible publications devoted to disseminating pictorial information by photographs. This industry quickly captivated the public and has continued to grow steadily to this day. Photography has often been called the democratic art, and the publishing boom in the 1870s foreshadowed our modern idea of photography for the masses.

◆

The first published Woodburytype print appeared in the January 1866 issue of the *Photographic News*, *A Mountain-Dew Girl: Killarney*, *(plate 21)* a genre-study photograph by H. P. Robinson, who was one of the leading proponents of art photography and author of *Pictorial Effect in Photography*, which also used a Woodburytype for its frontispiece *(plate 22)*.

Books like this, in every genre from photography texts to biographies and histories, used Woodburytypes for frontispieces or sometimes single illustrations *(plate 23-25)*. Many times the illustration didn't accompany the text but was a portrait of the author, a great way to add personality to an otherwise featureless book.

The most remarkable books containing Woodburytypes, however, were illustrated throughout. One of the first broad genres that captured the public's attention was the travel photograph. Then as now, people love seeing pictures of places they have never been. Woodbury capitalized on this during his residence in the South Pacific and the photographs he made with James Page. In 1875 he edited *Treasure Spots of the World*. This book is a prime example of the beauties and potential of the Woodburytype process. It contains twenty-eight photographs by Woodbury *(plates 7 and 8)* and other leading photographers of the day, including his friend John Thomson, with accompanying essays on each picture. This was the first model for publications like *National Geographic*.

> *The object of this, the first gift-book of its character, is to place before the public a selection of the most celebrated of the world's beauties and wonders. . . . The photographs are all taken by gentlemen whose names stand high in the photographic profession; and the proofs being printed in imperishable pigments by the Woodbury Process are thus guaranteed from fading or ever losing their brilliancy.*[70]

Photographs of places people had never seen were not always limited to exotic locales. Sometimes people didn't even know their own city and its inhabitants. John Thomson wanted to rectify this fact and spent years documenting the slums of Victorian London. His work culminated in the publication of the book *Street Life in London*.[71] One of the most extensive books of its kind from that period, the original edition contains thirty-eight photographs all printed in Woodburytype with essays contributed by Adolphe Smith, a prominent figure for the advocacy of the lower classes. It was first issued in serial form, then as a complete bound volume. Later, several smaller editions appeared, with an edited selection of the images.

In stark contrast to Woodbury's *Treasure Spots*, Thomson's *Street Life* gave the bourgeois an uncompromising vision of the lowest classes in Victorian London. Public attention had been called to the plight of the poor in one of the world's wealthiest cities by other works, such as *Oliver Twist*. Dickens realistically portrayed the seedy side of life through fictional characters, but Thompson's work was far from fiction. Here we see the seller of shell fish, the shoe-black and a stunning portrait of a "crawler," all real people in their everyday settings. Often seen but usually overlooked, here are these people in all their reality *(plates 26 and 27)*. The social realist photography of today owes a great debt to Thomson and his book in Woodburytype.[72]

> *Huddled together on the workhouse steps in Short's Gardens, those wrecks of humanity, the Crawlers of St. Giles's, may be seen both day and night seeking mutual warmth and mutual consolation in their extreme misery. As a rule, they are older women reduced by vice and poverty to that degree of wretchedness which destroys even the energy to beg.*[73]

A new social consciousness was emerging. The concern for greater involvement in society, partly due to the new affluence of the middle-classes, started extending into government. *Men of Mark* was a distinctive publication giving photographic portraits and biographic essays of the most important male figures in British society, from every field, but concentrating mostly in government and religion. All the portraits are fine examples printed in Woodburytype.

True democracy merged in art and politics when, for the first time, citizens could see and own real likenesses of their representatives.[74] A new transparency and openness came into government. The noble purpose of the publication to bring true likenesses of leading figures to all the people (as quoted above) gives us an indication of the power ascribed to the new ability to easily reproduce photographs.[75]

The easy printing of portraits helped propel the first mass political campaigning based upon personality. Carte-de-visite portraits became election tools and gave voters a glimpse of candidates they may never have seen in person. They are an ancestor to modern campaign posters and television ads. Other political figures not worried about campaigning, such as the royalty of the day, also used the carte-de-visite portrait to increase their visibility *(plate 28)*.

Portrait work was not limited to the documentary ideal, but was also appealing on a personal level. The strength of personality is nowhere more apparent than in entertainment. We can see the rise of true media celebrity in the theatre community that was a staple of middle and upper-class diversion in the late nineteenth century. The power of the Woodburytype was not lost on this industry and promoters were quick to put it to use. The allure of taking home a true likeness of one's favorite stage personality was inescapable. And for the first time, even if you couldn't get in to the theatre, by lack of luck or fortune, you could still gaze upon the face of your idol. Large and small, legendary and forgotten actors of the time are all captured forever in the Woodburytypes printed in many programs and journals of the time *(plates 14, 17 and 29)*.

Paul Nadar, son of the legendary Felix Nadar and well known photographer in his own right, was among the many people who photographed and published portraits of the French theatre community. He published *Paris Photographie* illustrated with photographic plates, occasionally using Woodburytypes *(plate 14)*.

Galerie Contemporaine, the French equivalent of *Men of Mark*, contains hundreds of portraits and short biographies of leading figures in French Society. It is the most extensive, continued use of the Woodburytype in a single journal, published in many parts over a period of about nine years (ca. 1876-1885).[76] Some of the issues contain military and political leaders and scientists, but the real concentration was on art—literary, performing and visual *(plates 29 and 30)*. All the issues with visual artists also have a plate of their work printed in Woodburytype *(plates 31 and 32)*. These beautiful volumes with their Woodburytype reproductions acted as a traveling gallery for the French sophisticate.[77]

The visual arts community quickly caught on to the merits of the Woodburytype because it possessed two great advantages over previous illustration methods that the art world was eagerly awaiting: its ability to render true shades of gray, and to duplicate an existing image through the use of a photographic negative *(plates 32-34)*. Now a painting could be photographed and printed in Woodburytype. This was the first time paintings did not have to be translated into the medium of engraving or woodcut with lines or dots to simulate the subtle chiaroscuro. Artists were eager to have their work, not the interpretation of a copyist, seen by the general public.

> *Thanks to the perfection attained at the present time by photography, the visitor of to-day to the Art Galleries of England and the Continent can gratify his tastes in a manner unknown to his ancestors. In years not very remote, an expensive engraving of a few of the most important pictures was all that could be obtained to remind those of artistic instincts of what they had seen and enjoyed among the picture shows of the world. But the photographer has altered all that.*[78]

The *Photographic Art Journal*, which began publication in 1870, was one of the first large journals to make use of the Woodburytype.[79] Their stated aim was to reproduce the world's finest art through the new photomechanical processes.

> *Our illustrations . . . will be as a rule such as have a bearing upon Art rather than Science or Mechanical applications. The title which we have adopted sufficiently points to the class of subjects which the reader may expect to find illustrated in this Journal. These we shall confine to views and studies from nature, artistically selected; to reproductions of fine works of art by eminent artists; portraits; and examples of the value of the improved processes of photography in superseding the costly, tedious, and often inaccurate work of the engraver.*[80]

The first issue was devoted to the Woodburytype and contains a number of fine illustrations of paintings, photographs and other art work printed in Woodburytype. The editors valued the process to such an extent that they printed an article describing it, going into great detail about the firm Goupil, which was contracted to supply all the Woodburytypes for the journal.

◆

Just as important as the publications that used Woodburytypes were the shops that printed them. Goupil was by far the largest supplier of prints in the early years. They were one of the first companies to buy the patent rights, in late 1866, and the first to perfect the process while it was still in its developmental stages.[81] They expanded their existing facility in Asnieres to accommodate the Woodburytype, and the firm's manager, Roussellon, worked directly with Woodbury as they ironed out the process for commercial operation.[82] The *Photographic Art Journal* reported:

> *After the failure of no less than three Companies in England for carrying out this invention commercially, the well-known publishing firm of MM. Goupil, of Paris, have at length succeeded in the attempt. They have erected a large manufactory at Asnieres, near Paris, have put up costly machinery for working the process on a large scale, and have every reason to look for a great success. . . . Works which now take months and years of the most skilled labour to engrave, and which cost hundreds and thousands of pounds, can by this marvelous process be executed far more truthfully in a few hours, and at a cost comparatively nominal.*[83]

Adolph Goupil started as an art dealer.[84] When his firm bought the rights to the Woodburytype for 150,000 francs they were already known for reproductions using traditional print methods.[85] It was a logical step for them to get involved with this avant-garde technique. They continued producing much the same inventory for both individual sale and inclusion in publications. By the time the shop was established, they had installed thirty presses, which was not enough to keep up with the demand.[86] Even though they were printing from two plates on a single press, they still had to install more equipment to fill orders.[87] But the Franco-Prussian war hit Paris in 1870 and badly damaged the facility. Because of their early success, they were quick to rebuild, dedicating an entire floor to the Woodburytype and adding more personnel to run the process.

Another invention made practical around this time was the electric light. Relying on the inconsistency of light from the sun to expose the relief film became a challenge to rapid production. The electric light provided a means to keep work going at any time of the day or year, irrespective of sunshine. Goupil took advantage of this and installed one of the first electric lights (and the generating plant to run it) in a photographic establishment.[88] By 1873 operations were restored and surpassed the original line with thirty-six presses running. "Three powerful apparatus, together with a personnel of sixty, are employed for the daily work; one steam engine, a hydraulic press, and an electric battery are also used."[89]

Although the *American Journal of Photography* reprinted Woodbury's first English article on the Woodburytype in May 1865 and Woodbury secured U.S. patent rights in 1866, real news of the Woodburytype didn't come to America until 1869 when Edward Wilson, editor of the *Philadelphia Photographer*, demonstrated the process at the *Philadelphia Photographic Society*.[90] In the audience that day was the photographic inventor, printer and entrepreneur from Chicago, John Carbutt. He was so taken by the beauty and possibilities of the process that within the year he had acquired the patent rights, traveled to England to be taught by Woodbury, moved his print shop from Chicago to Philadelphia and started working the Woodburytype.

Carbutt's previous experience with silver prints for books such as *Leading Men of Chicago* qualified him for the sole ownership of patent rights in the United States. He hoped that the Woodburytype would speed and simplify his operations. He turned out quality prints and lantern slides under the firm American Photo-Relief Printing Company (APRPC) for about five years. The new facility, the only one in America at the time, was quite large, measuring three stories. He won a silver medal from the American Institute in 1873 for Woodbury prints on paper and glass, and in 1874 the Franklin Institute awarded him their silver medal for Woodbury lantern slides.[91]

But by the mid 1870s, being the entrepreneur that he was, Carbutt began using other, faster mechanical processes. There is some inference that Carbutt continually experienced problems with the weather in America, particularly the drastic changes in humidity.[92] The difference in climate between the mostly cool European countries and America may have caused solubility, setting and cracking problems with the gelatin. One remedy for this was to thin the solutions. In a print this would produce a less glossy surface as the percentage of gelatin decreased and indeed most of Carbutt's Woodburytypes exhibit a matte appearance.

Whatever the reasons, Carbutt sold his interest in APRPC to a firm that moved the facilities to Boston, where they were destroyed by the great fire, effectively ending the Woodburytype in America.[93]

The fate of Woodbury's process in his home country wasn't as simple as in France or America, partly because he had trouble with the patent rights and financing. There are many confusing and contradictory reports in the press and it is difficult to separate true fact from Victorian fact. What we do know is that sometime before 1868 a shop was set up in Manchester (Woodbury's hometown) but they were unable to produce anything in quantity. Then in 1868, the French photographer so well known for the carte-de-visite, A. A. E. Disderi, procured the English patent rights.[94] He established a shop in Brompton, London under the name the Woodbury Permanent Photographic Printing Company (WPPPC).[95] Under his direction, the firm appears to have suffered financial trouble, failed to make payment for the patent and abandoned the Woodburytype.[96] The *Photographic News* reported:

> *Fine examples of the work begins to be seen in retail establishments, and a full tide of prosperity seems to be promised, when suddenly Mr. Disderi is reported as "wanting," and shortly his solicitor applies to the Court of Chancery for a liquidator, and states that no money at all has been contributed by the shareholders and directors . . . that no less than £40,000 has been supplied by a French capitalist to the company in London to initiate and carry on its business, but that no satisfactory answer can be given as to where the money has gone.*[97]

After this trouble the firm's entire stock was liquidated and sold to the lithographer Vincent Brooks.[98] Woodbury took over the management of these works sometime in 1869 and apparently held stock in the company for at least a few years. By 1872, when the patent expired, Woodbury was no longer part of the company. The last known Woodbury Company was taken over by Eyre & Spottiswoode, who continued to use the Woodburytype at least until 1906 for lantern slides and probably for postcards as well.[99]

Many other firms in Europe worked the process at various times, including Adolph Braun, who bought the rights and established his shop in Dornach, Alsace.[100] There was also at least one shop set up in Australia.[101] J. Waterhouse, head of the government printing office in India—the largest printing office in the world at the time—also probably set up a Woodburytype facility as he lectured many times on the merits of the process.[102] But no prints have surfaced from these areas. (See appendix for list of firms licensed by Woodbury.)

◆

The rapid growth of the Woodburytype in the short time between 1869 and 1872—including the establishment of shops around the world and the sheer volume of material issued by them—indicates how far the Woodburytype penetrated into society. From a tiny number of books with photographs in the early 1860s to a plethora of publications, including magazines, by 1870, the world was a very different place. The ability to produce so many prints inexpensively made possible the production of books like *Street Life in London* and the *Photographic Art Journal*. With traditional or silver prints, it would have been prohibitively expensive to issue these publications in any quantity.

Cost was extremely important for firms investing in new print technologies. The failure of the English firms and the amount of capital laid out by Goupil indicate what was required to start a venture or convert an existing shop. To establish a Woodburytype facility, additional personnel were needed, but most of the production line could be manned with printers trained in the traditional techniques. Only the plate maker needed a thorough working knowledge of photographic chemistry.

New and rather expensive equipment was also needed, such as a hydraulic press to make the printing plates, the printing presses and other miscellaneous pieces. A fully equipped shop could be established for around £3,750 (1870).[103] But with the increase in production, a healthy firm would easily recoup these expenses by supplying large editions.

Just how many prints could be produced was the subject of much debate in the photographic press. Because of varying claims, an accurate estimate is difficult. Woodbury's reports seem inflated, at times double or triple what others were producing.[104] Reports in the press ranged between 200 and 2,000 prints per plate. With multiple plates being used (the conventional tables held six presses), an edition size of 5,000 was perfectly feasible.

All five of the photographic journals, *Photographic News*, *British Journal of Photography*, *Philadelphia Photographer*, *Photographic World*, and *Photo Mittheilungen*, inserted Woodburytypes with an issue. The German photographer, Vogel, stated Woodbury had made 1,500 prints for the *Mittheilungen*. And Wilson stated 5,000 prints were made for the *Photographic World*. Wilson also said the publisher Frank

Leslie had printed "75,000 copies of one picture for gratuitous distribution with his periodicals." That is the largest cited edition size known for a single image.[105]

The number of prints versus the overhead, of course, determined the sales price of the final prints. This was particularly important for publications, but is more easily reflected in prints for individual sale. The Woodburytype, because of its increased edition size, substantially lowered retail prices.

> *Mr. Woodbury informed me that they print one hundred* cartes-de-visite *or quarter-plate proofs for twenty shillings for the first hundred, and nine shillings for every succeeding hundred—the greater cost of the first hundred being occasioned by the intaglio, which costs twelve shillings. Prints 5 x 4 can be supplied for £6 per 1,000. Stereoscopic prints about thirty shillings per first hundred, and so on in equally reasonable proportion.*[106]

Another report of a farthing a copy is about the same as Woodbury's 5 x 4 cost.[107] The *Photographic News* reported with the issue of the first published Woodburytype that its cost was "one-fifth of a penny each."[108] This was far cheaper than other photomechanical processes. At this time, the photogravure, for example, cost about as much to produce one printing plate as an entire edition of Woodburytype prints.[109] The same was true for silver prints:

> *There are few persons who have not seen and admired the stereoscopic glass transparencies of M. Ferrier, which even at the high price of six shillings each command an enormous sale; such slides can now be printed by the Woodbury process, and sold with profit at a third of that price.*[110]

Even though the Woodburytype was the most profitable process at the time, it did have its disadvantages. It is a very laborious process, leading the *Photographic News* to write:

> *The labour employed must be trained to manipulative skill, and half the term of the patent may expire before success is sufficiently assured to attract capital, by which the process may be developed into a profitable industry.*[111]

Part of the reason for the delay in producing good prints was that the necessary machinery had to be specially made. There were no hydraulic or printing presses extant that were appropriate for the process. Goupil and other firms experimented with different models and had several custom-built until they found presses that worked.

To most people, however, these disadvantages were outweighed by the beauty of the prints. The German photographer Grasshoff, discussing the merits of the Woodburytype before the Society for the Advancement of Photography, Berlin, said:

> *I was very soon convinced that none of the many manipulations can be spared, and through the astonishingly beautiful results we are easily recompensed for all the trouble of manipulation.*[112]

PLATES

Photography credit: Images for plates Frontispiece, 1-7, 10, 11, 18, 21, 22, 34, 36-49 and 52-59 courtesy of the National Museum of Photography, Film and Television, Bradford, England. Images for plates 8, 9, 13, 14, 23-32, 35 and 51 courtesy of the Department of Photographs, J. Paul Getty Museum, Los Angeles. All other images provided by the author.

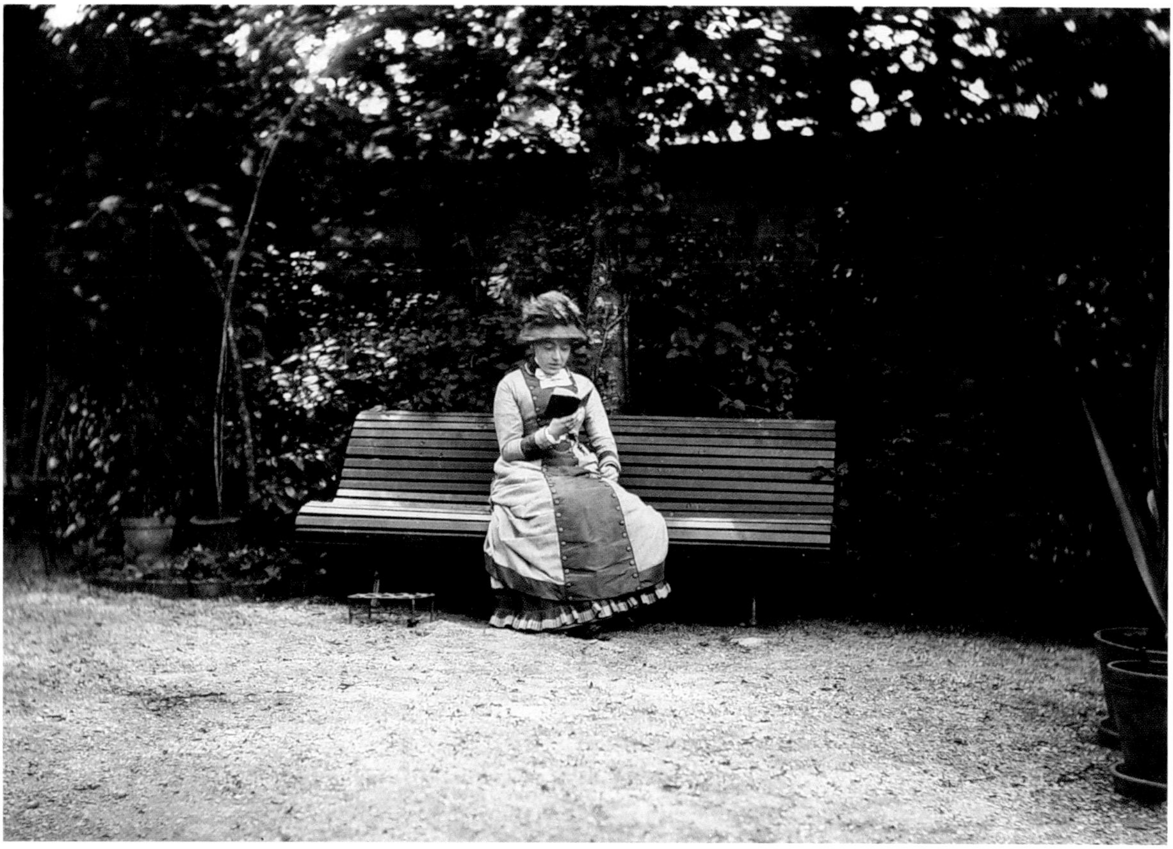

Plate F-1. Woodburytype of a woman in park seated upon a slatted wooden bench, photograph by C.H. Auger, 1875.

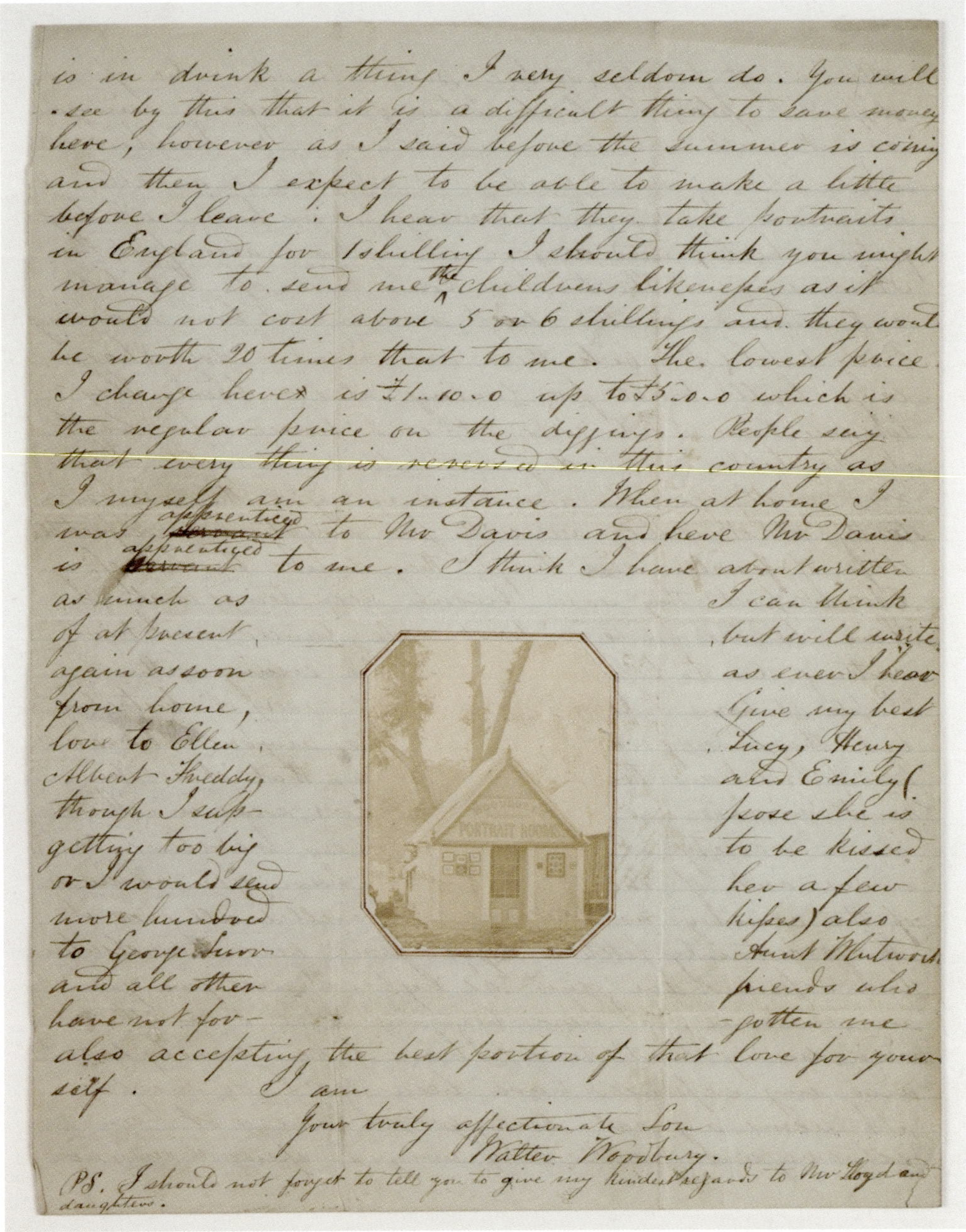

is in drink a thing I very seldom do. You will see by this that it is a difficult thing to save money here, however as I said before the summer is coming and then I expect to be able to make a little before I leave. I hear that they take portraits in England for 1 shilling I should think you might manage to send me the childrens likenesses as it would not cost above 5 or 6 shillings and they would be worth 20 times that to me. The lowest price I charge here is £1.10.0 up to £5.0.0 which is the regular price on the diggings. People say that every thing is reversed in this country as I myself am an instance. When at home I was apprenticed to Mr Davis and here Mr Davis is apprenticed to me. I think I have about written as much as I can think of at present, but will write again as soon as ever I hear from home, Give my best love to Ellen, Lucy, Henry Albert Freddy and Emily (though I suppose she is getting too big to be kissed or I would send her a few more hundred kisses) also to George Snow Aunt Whitworth and all other friends who have not forgotten me also accepting the best portion of that love for yourself. I am

Your truly affectionate Son
Walter Woodbury.

P.S. I should not forget to tell you to give my kindest regards to Mr Lloyd and daughters.

Plate 1. Letter from Woodbury in Australia to his mother in England, August 23, 1856, with a photograph of his studio.

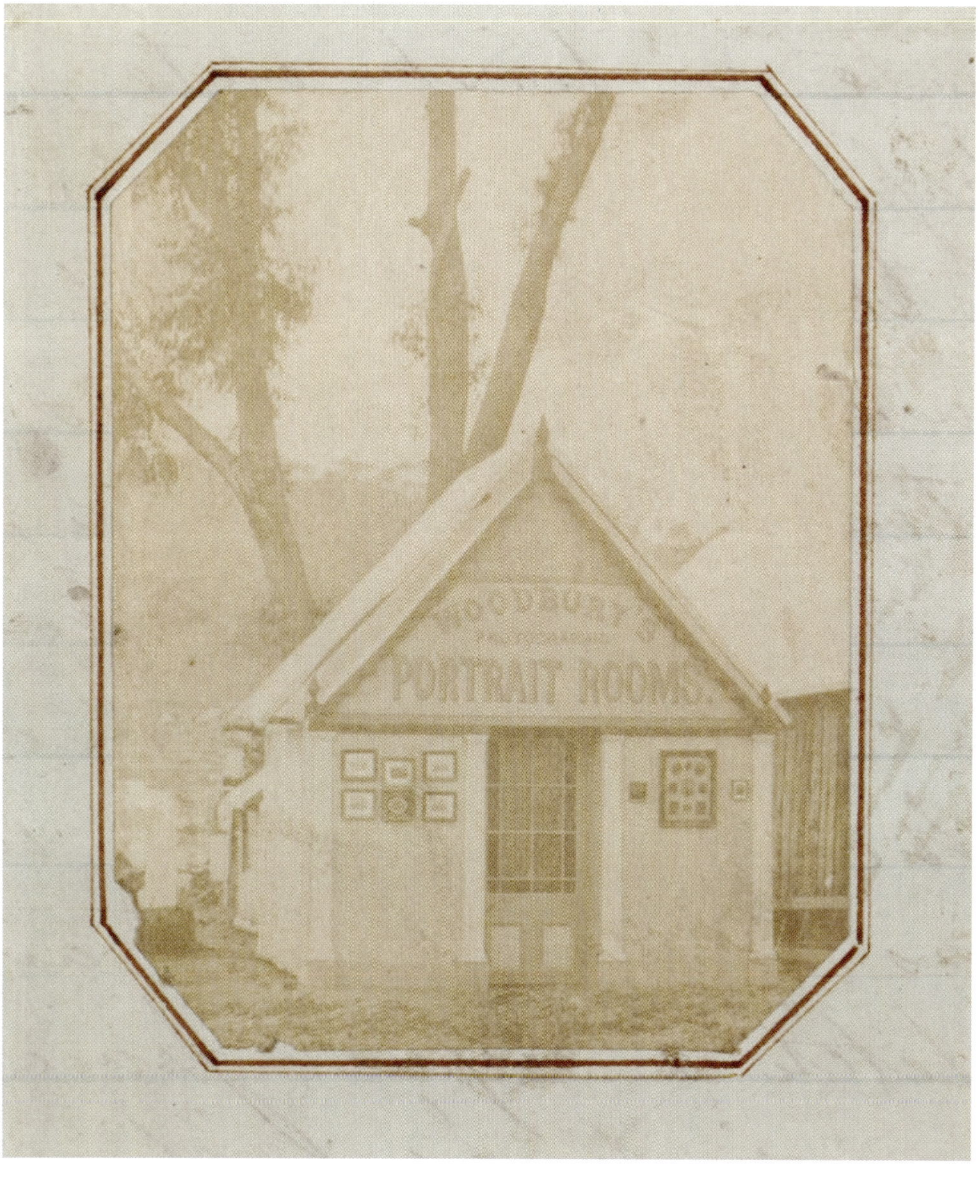

Plate 2. [Walter B. Woodbury], Woodbury's studio outside Melbourne Australia, 1856.

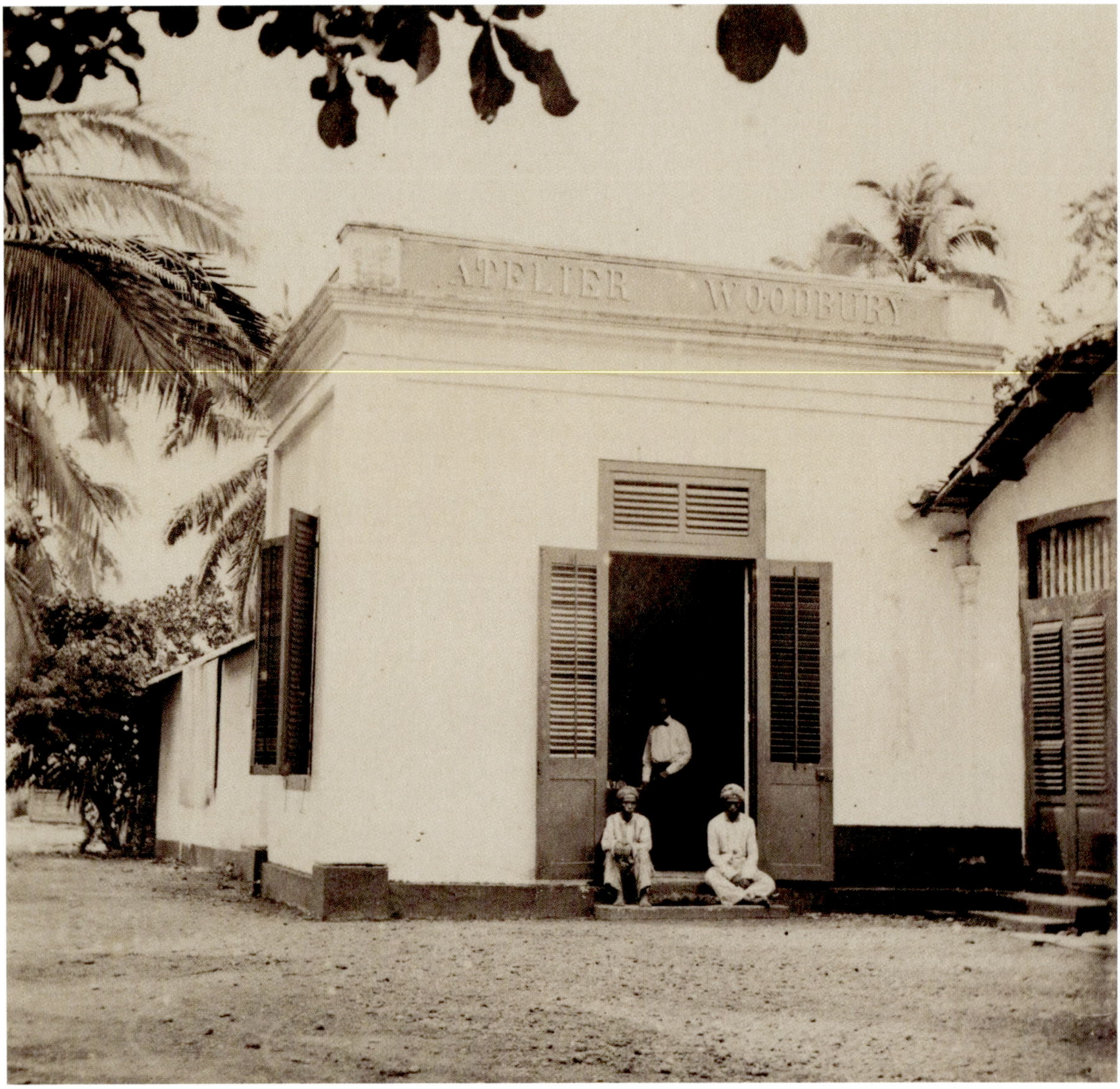

Plate 3. [Walter B. Woodbury], Atelier Woodbury, Java, [1861].

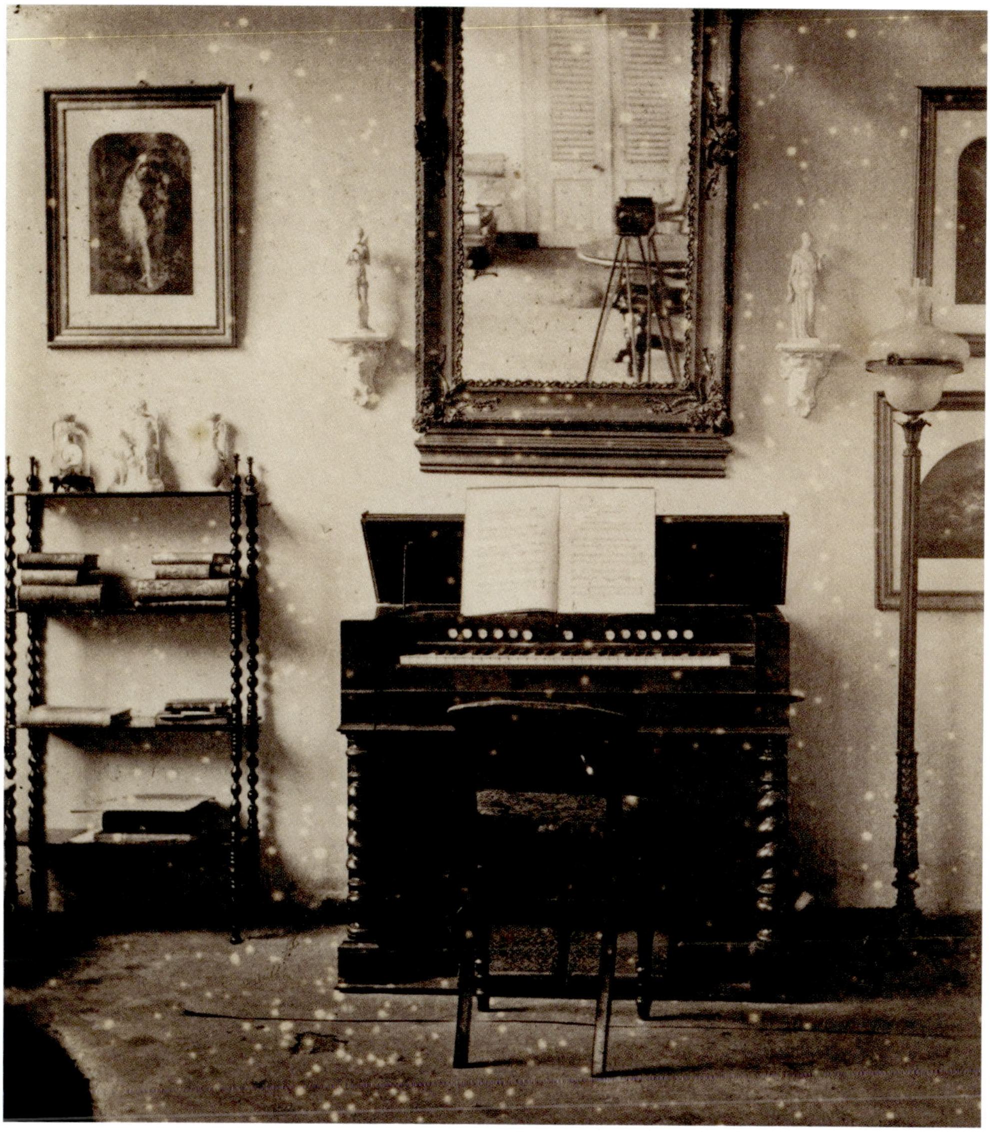

Plate 4. [Walter B. Woodbury], Interior Woodbury & Page Studio, Java, [1861].

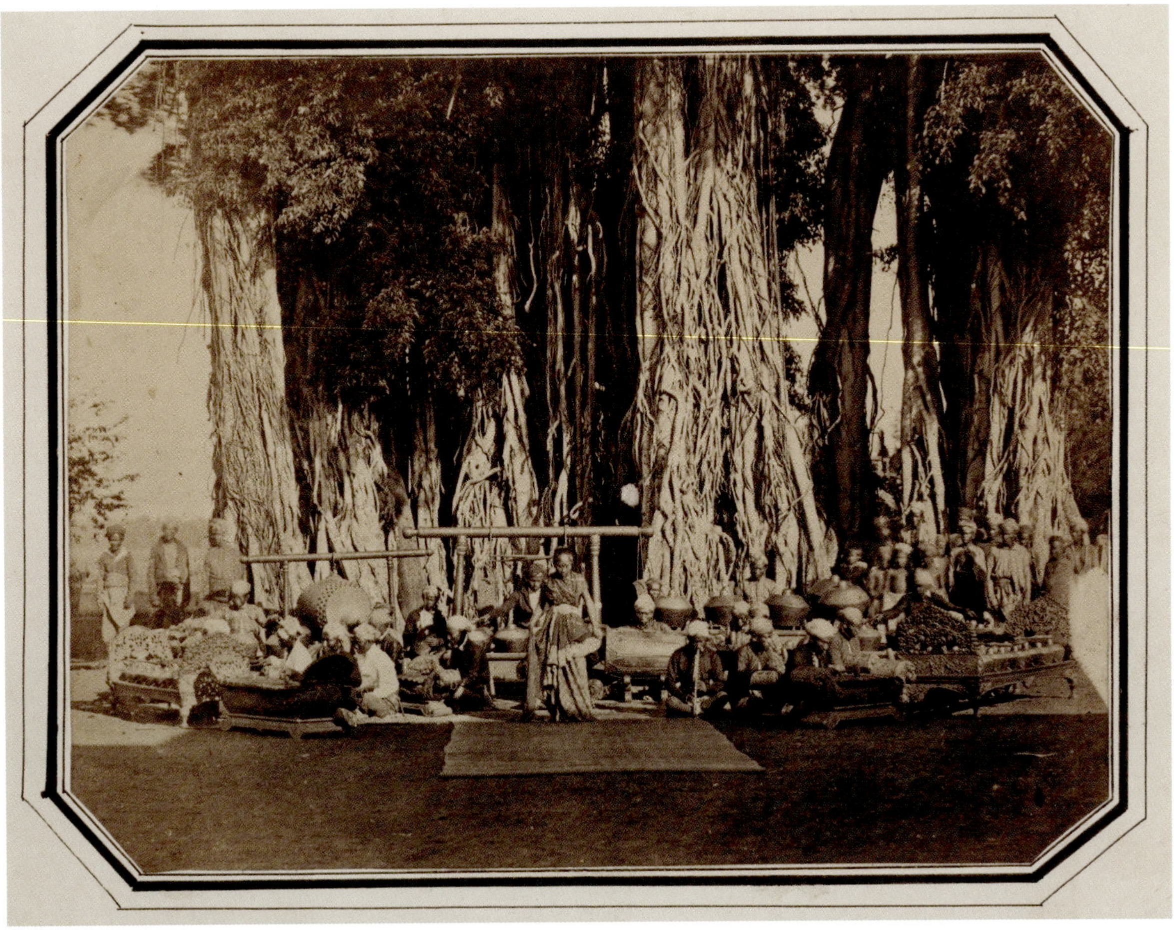

Plate 5. Walter B. Woodbury, Java, Musicians, [1861].

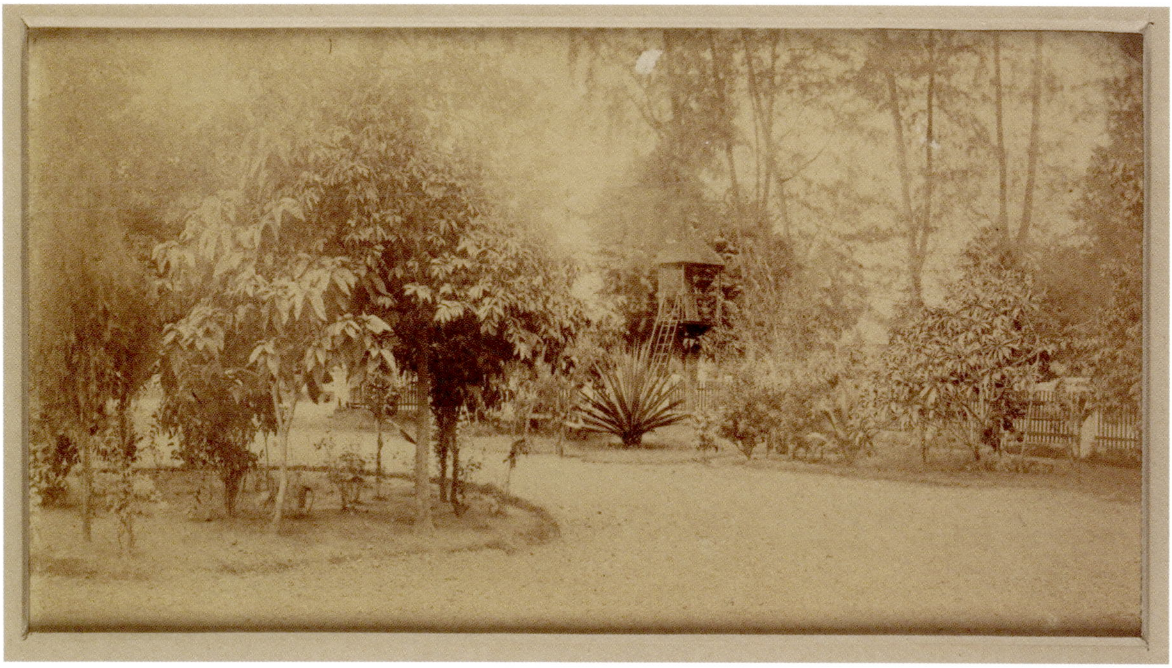

Plate 6. Walter B. Woodbury, Panorama, Java, [1861].

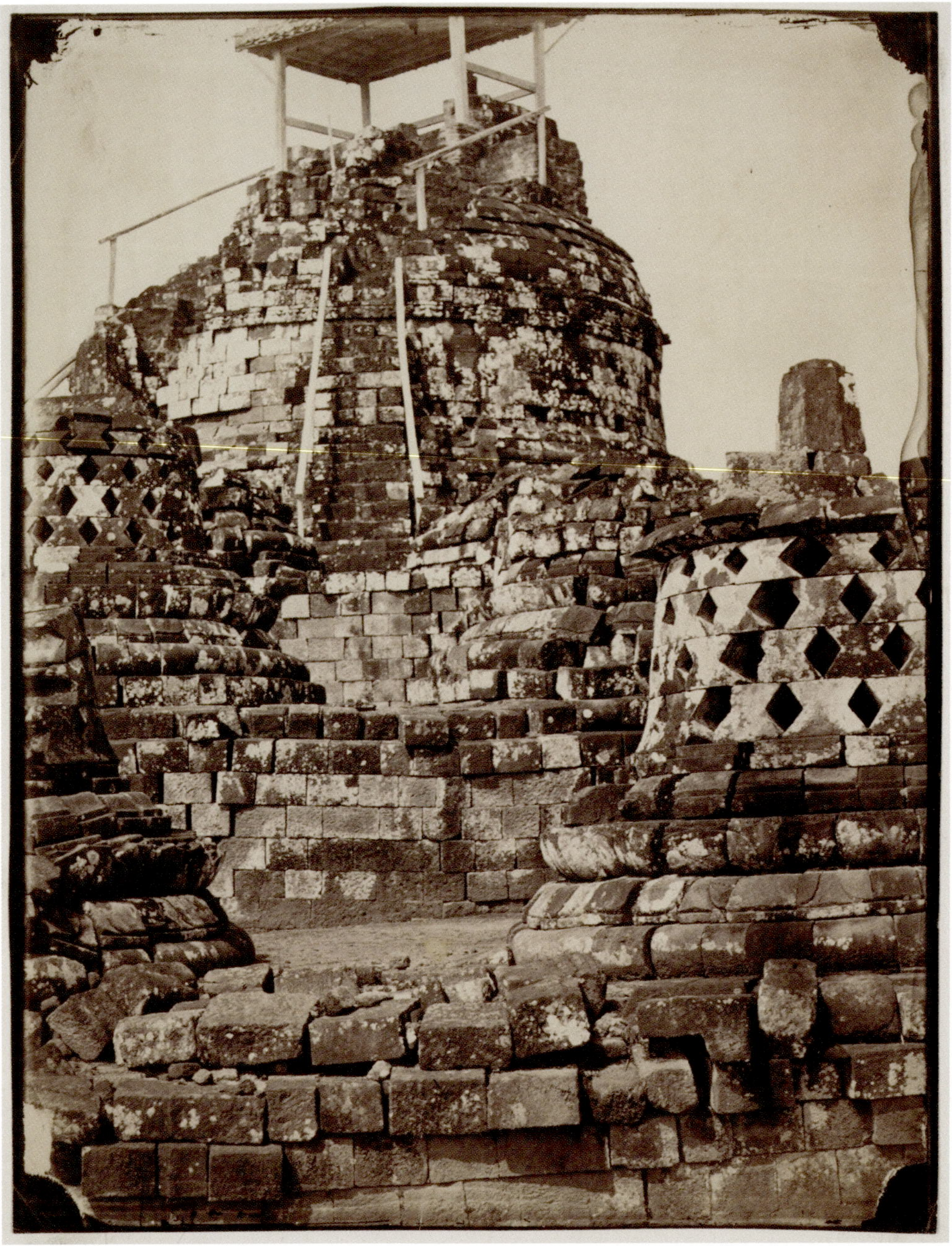

Plate 7. Walter B. Woodbury, Temple, Java, [1861].

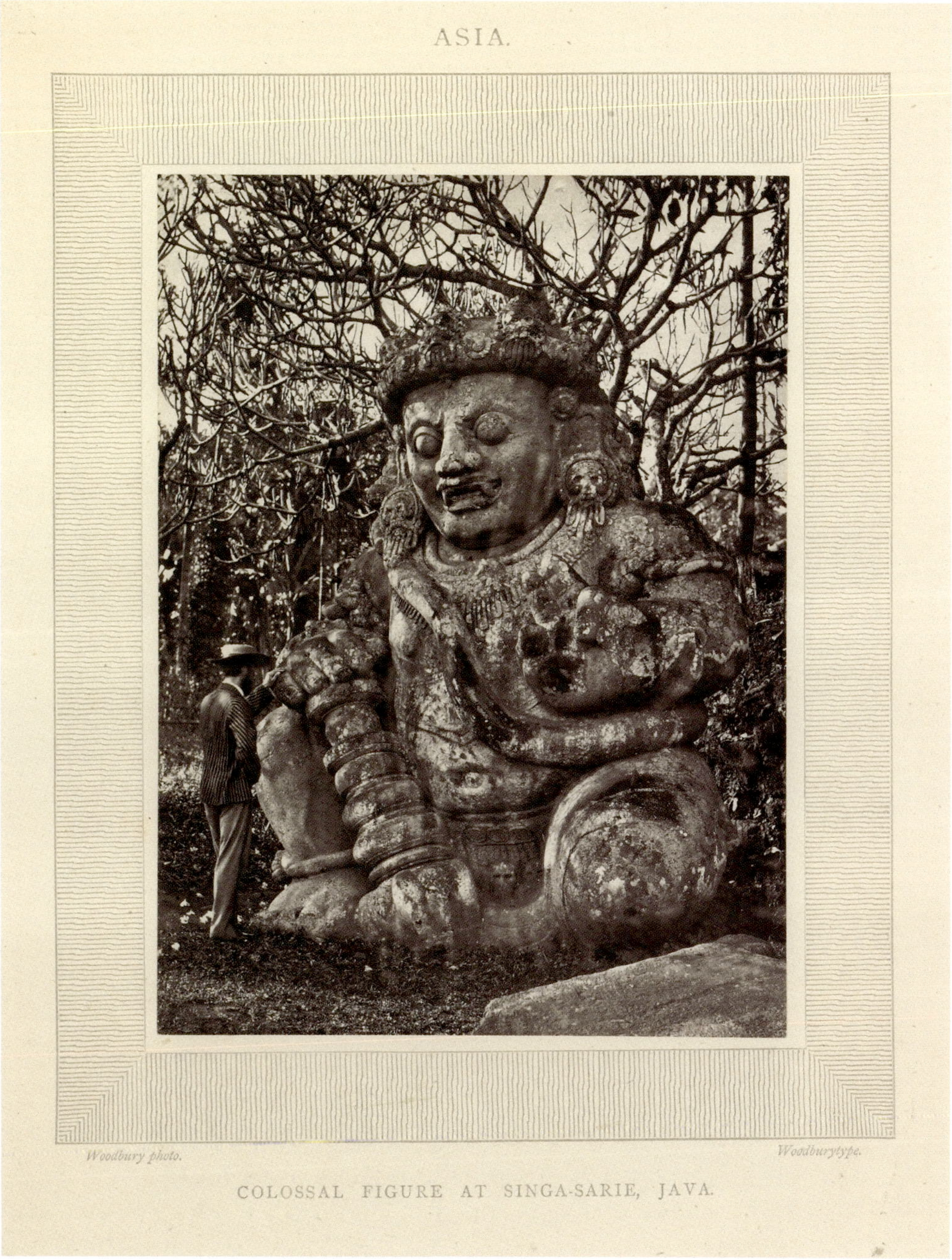

Plate 8. Walter B. Woodbury, Colossal Figure at Singa-Sarie, Java, from Treasure Spots of the World, *1872, negative made [1861].*

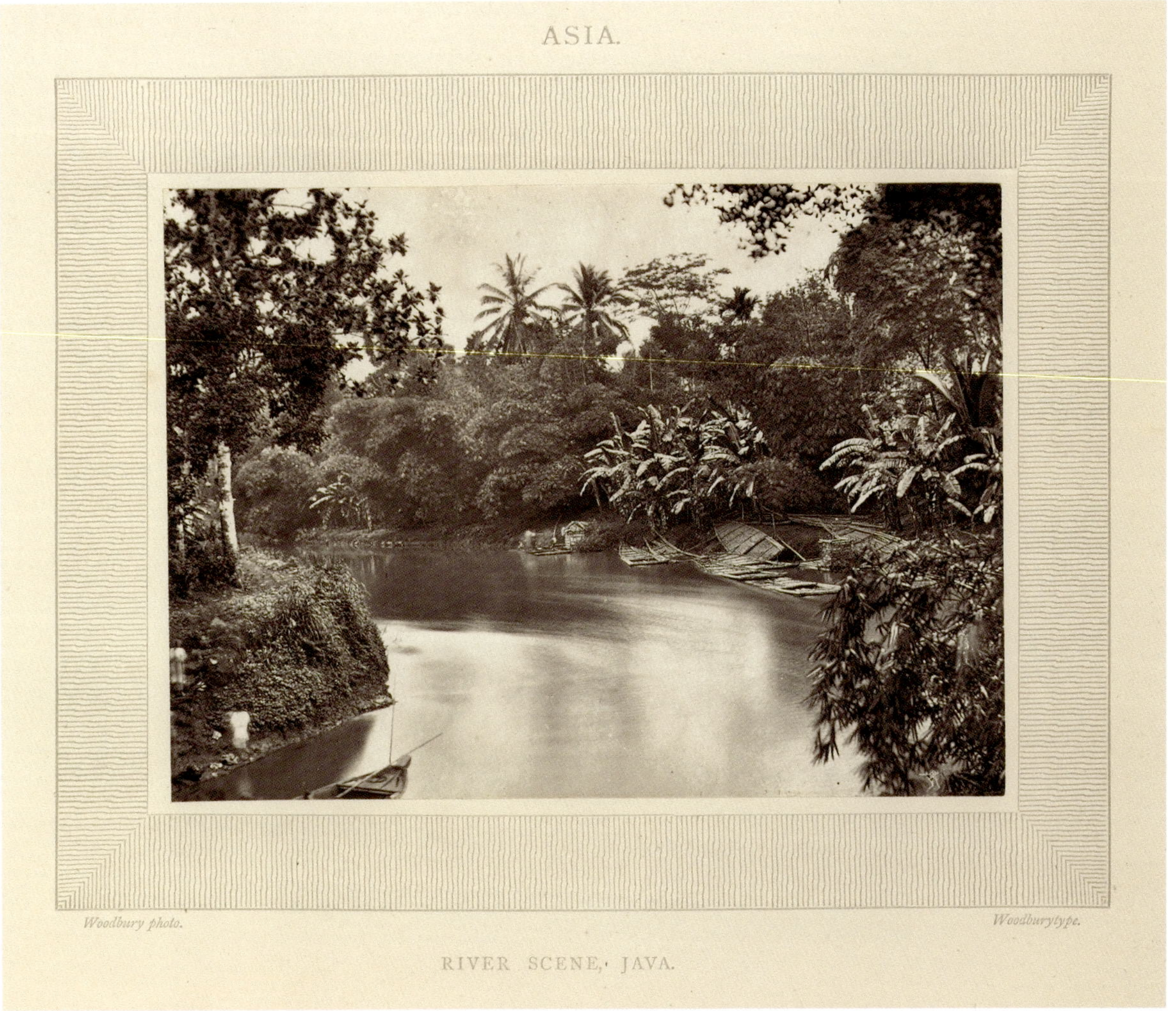

Plate 9. Walter B. Woodbury, River Scene, Java, *from* Treasure Spots of the World, *1872, negative made [1861].*

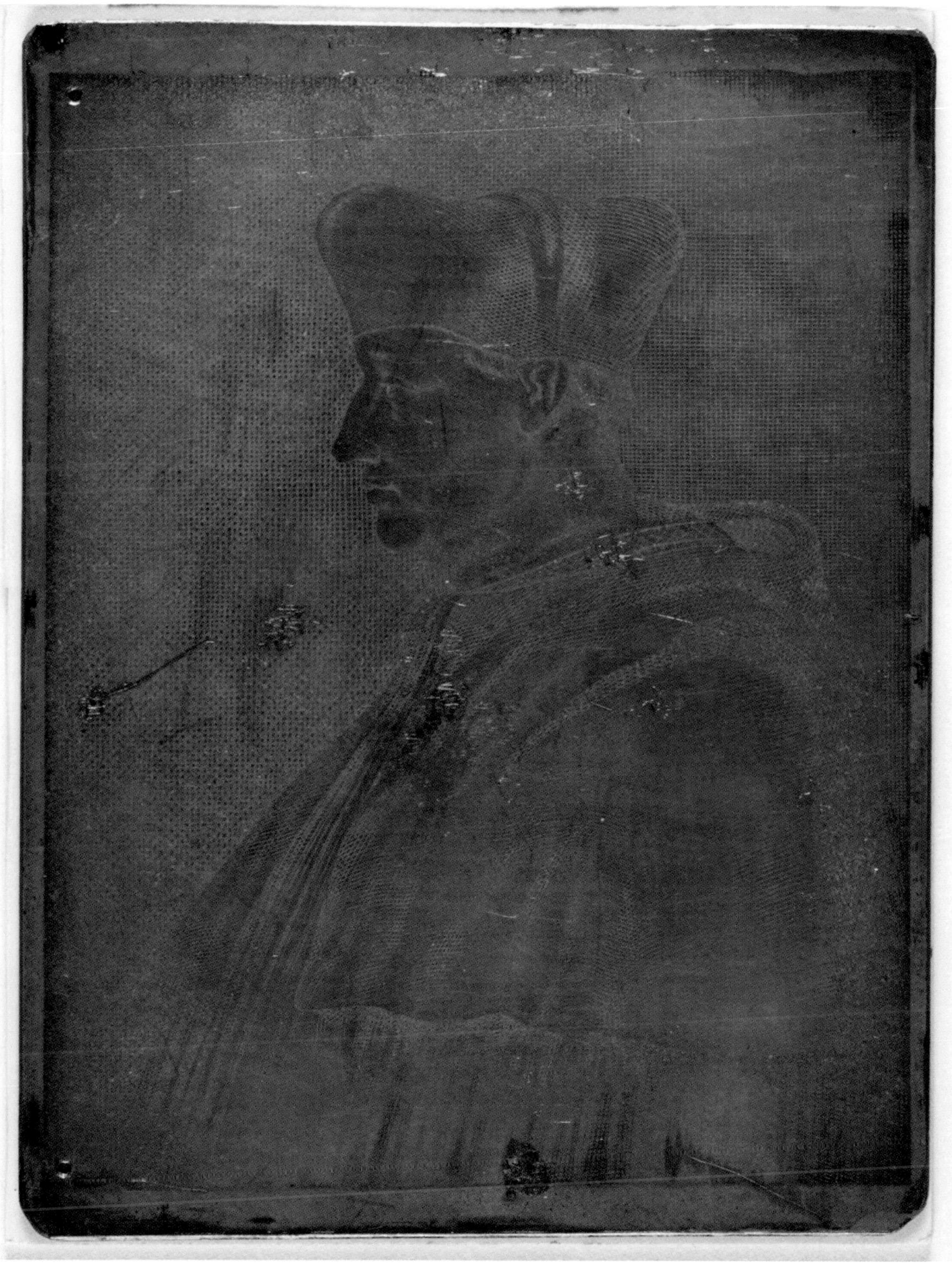

Plate 10. Joseph Nicephore Niepce, Portrait of Cardinal d'Amboise, *1827. Heliograph on pewter (see plate 11). One of the oldest surviving photographically produced images and the first photomechanical plate.*

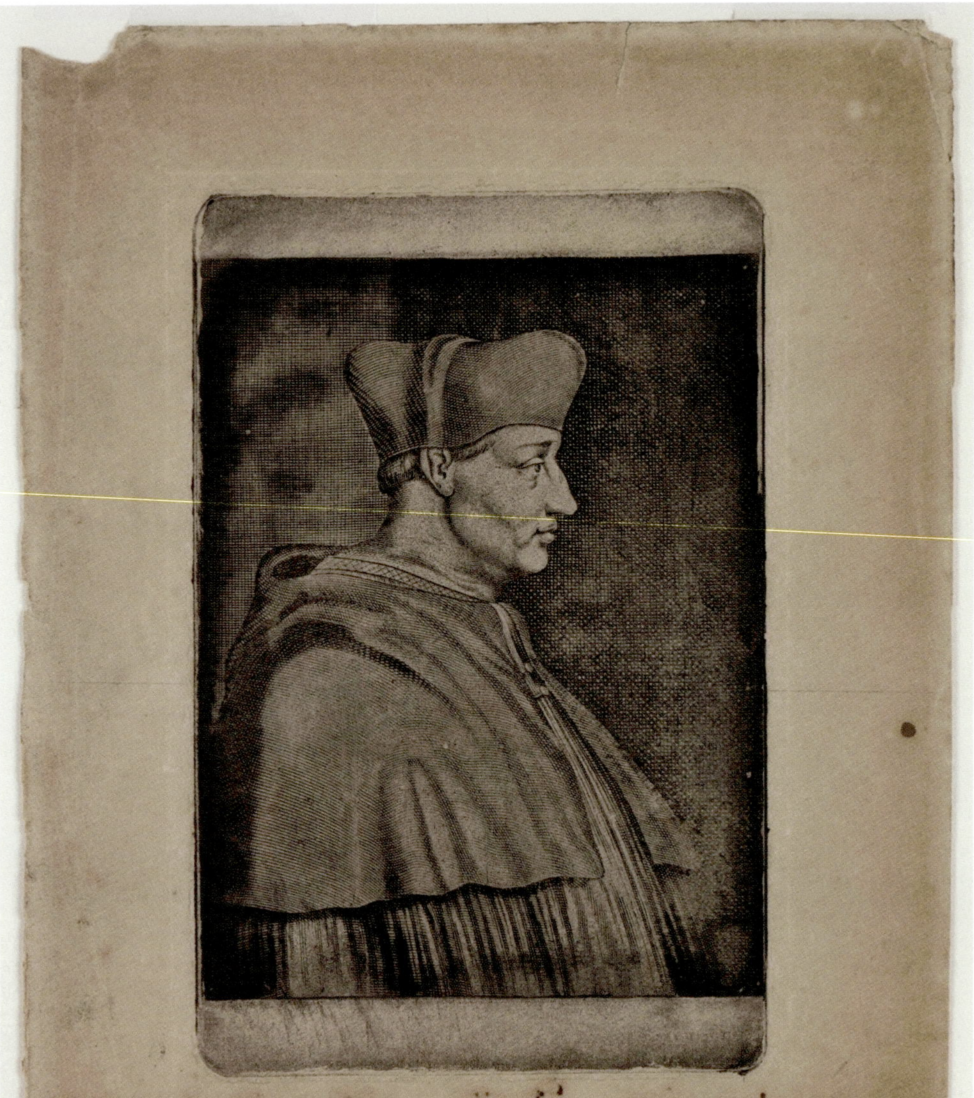

Plate 11. Joseph Nicephore Niepce, Portrait of Cardinal d'Amboise, *1827. Ink impression on paper from a heliograph on pewter (see plate 10). Possibly the first photomechanical print.*

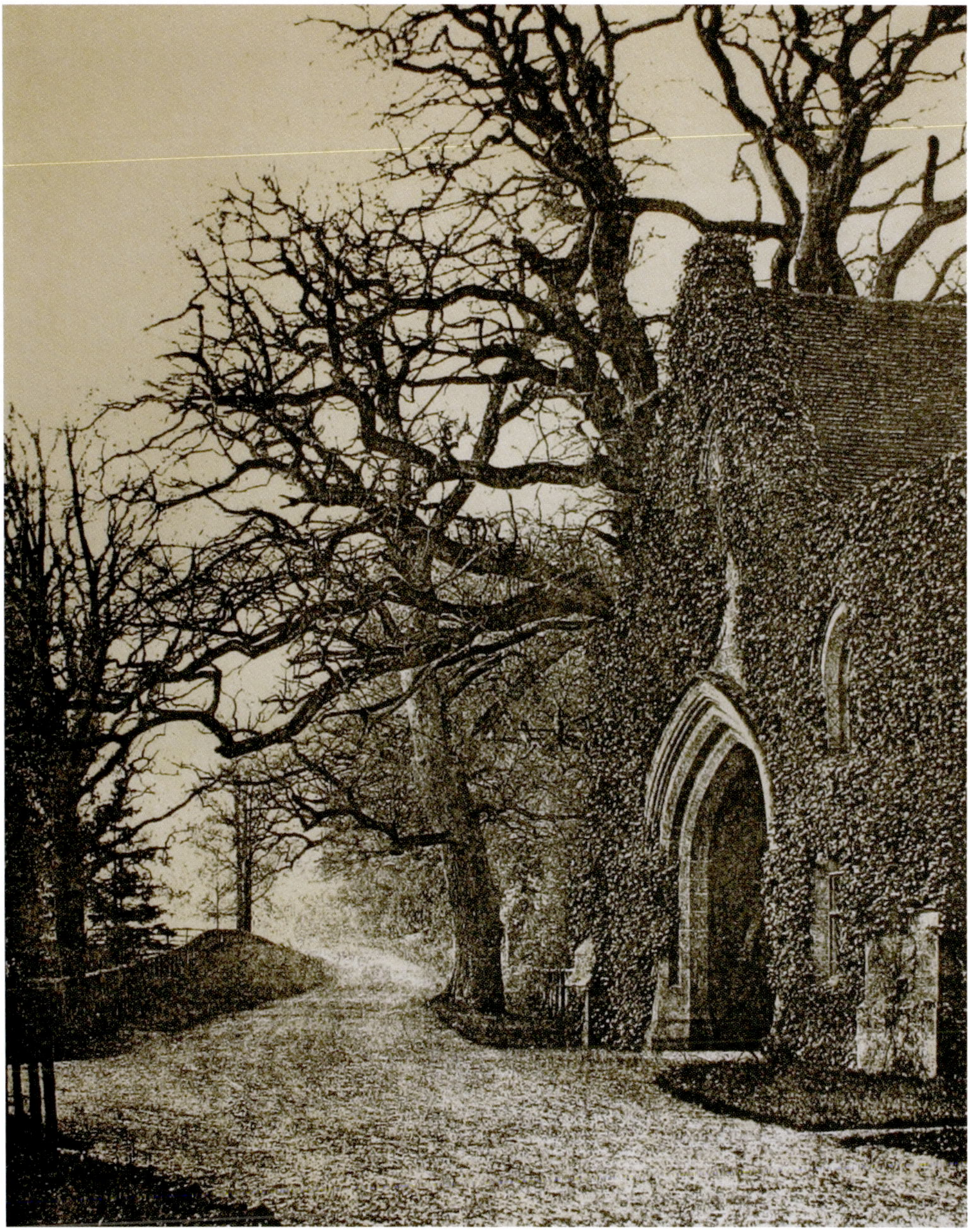

Plate 12. Photo-lithograph published in the British Journal of Photography, *April 27, 1866, from a negative by Dr. Ebbage. Printed by Messrs. Bullock Brother's patent process. This print represents the state-of-the-art in photolithography at the time of its publication. Notice the extreme granularity and lack of middle values compared with the Woodburytype at plate 21, published earlier the same year.*

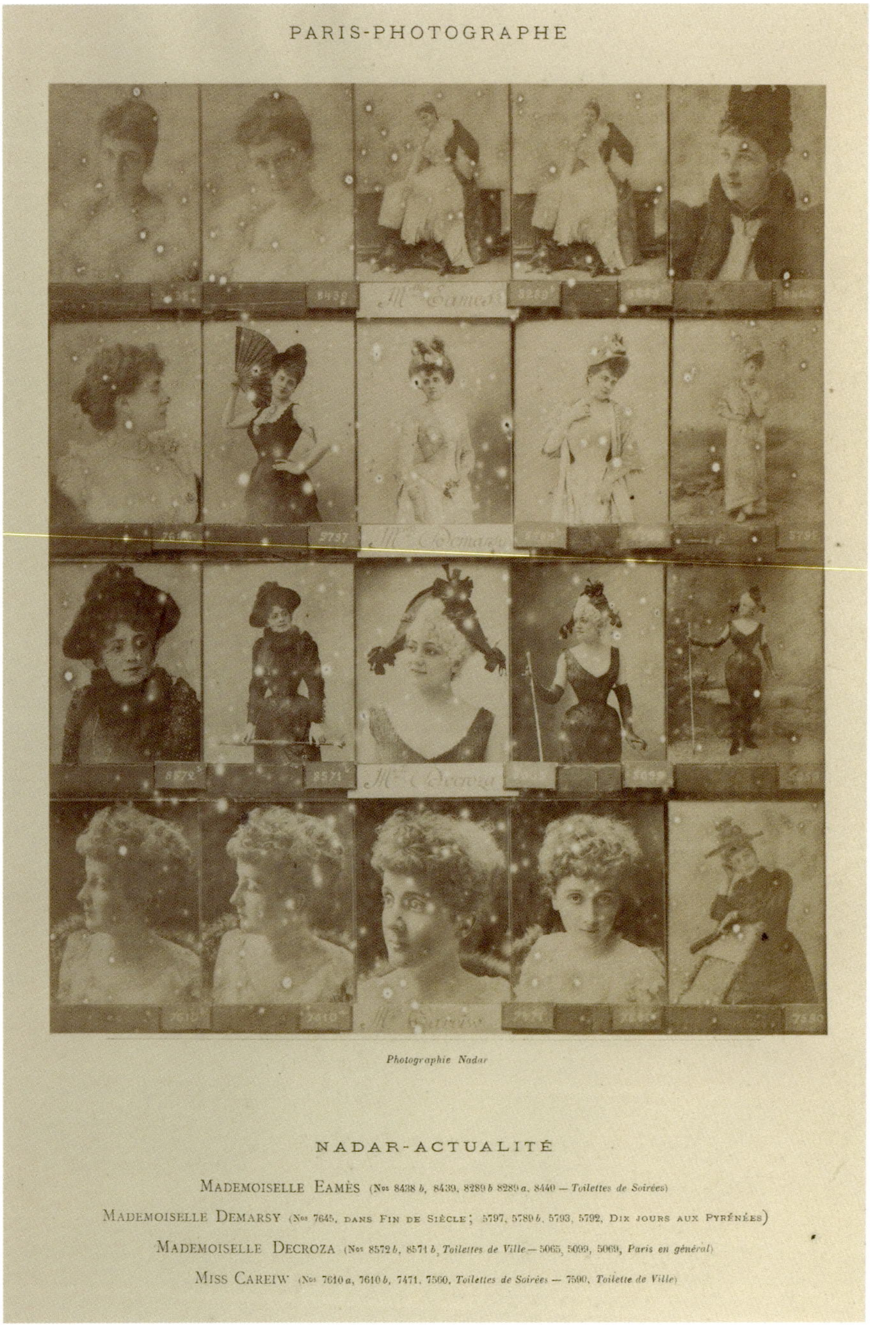

PARIS-PHOTOGRAPHE

Photographie Nadar

NADAR-ACTUALITÉ

MADEMOISELLE EAMÈS (Nos 8438 *b*, 8439, 8289 *b* 8289 *a*, 8440 — *Toilettes de Soirées*)

MADEMOISELLE DEMARSY (Nos 7645, DANS FIN DE SIÈCLE; 5797, 5789 *b*, 5793, 5792, DIX JOURS AUX PYRÉNÉES)

MADEMOISELLE DECROZA (Nos 8572 *b*, 8571 *b*, *Toilettes de Ville* — 5065, 5099, 5069, *Paris en général*)

MISS CAREIW (Nos 7610 *a*, 7610 *b*, 7471, 7560, *Toilettes de Soirées* — 7590, *Toilette de Ville*)

Plate 13. Paul Nadar, Vignettes of four actresses, from Paris Photographie, *no. 3, June 1891. This tipped-in albumen print shows all the typical signs of aging characteristic of poorly processed photographs from this time period. Notice the darkening at the edges, the fading in the center and the spots throughout, compared with the Woodburytype at plate 14 published the previous month in the same journal.*

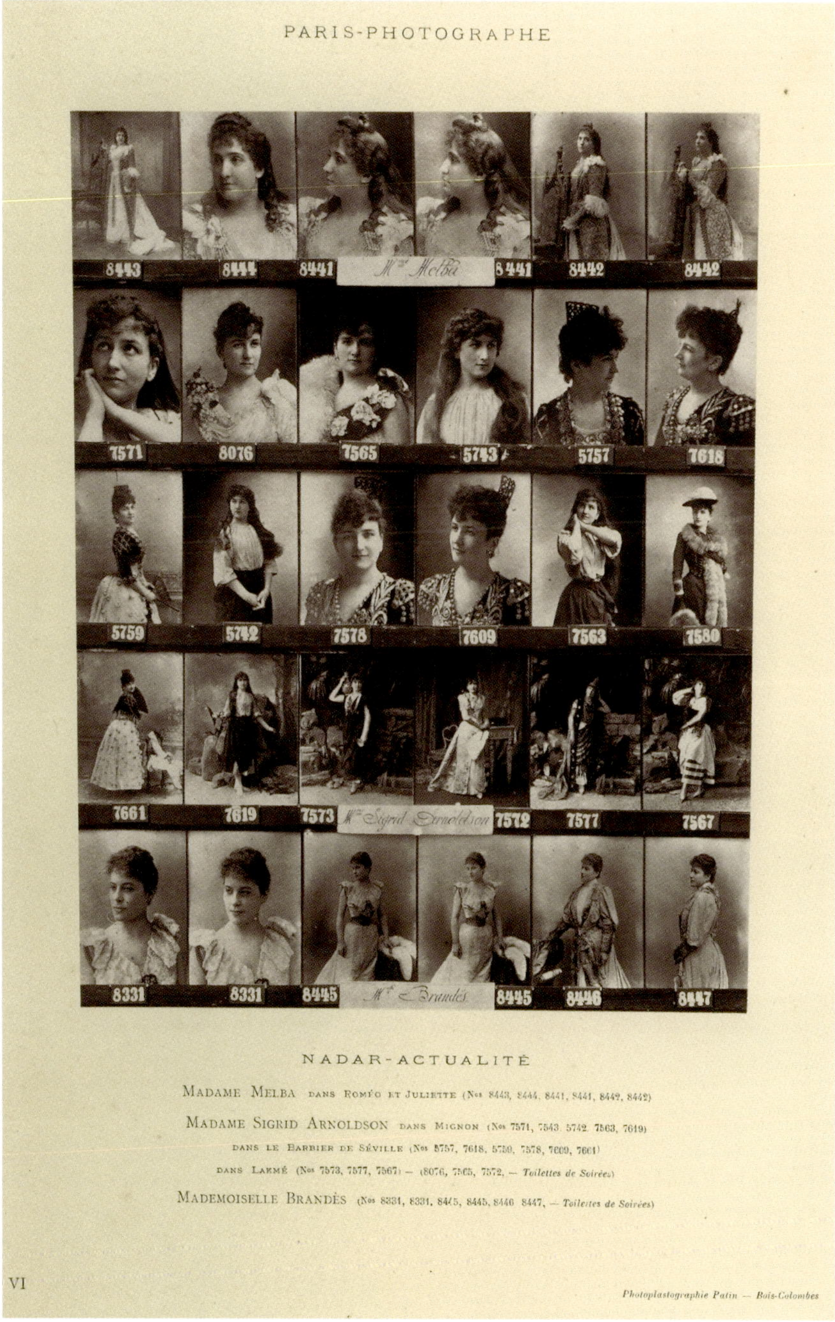

PARIS-PHOTOGRAPHE

NADAR-ACTUALITÉ

MADAME MELBA DANS ROMÉO ET JULIETTE (Nos 8443, 8444, 8441, 8441, 8442, 8442)

MADAME SIGRID ARNOLDSON DANS MIGNON (Nos 7571, 7543 5742 7563, 7619)

DANS LE BARBIER DE SÉVILLE (Nos 5757, 7618, 5759, 7578, 7609, 7661)

DANS LAKMÉ (Nos 7573, 7577, 7567) — (8076, 7565, 7572, — *Toilettes de Soirées*)

MADEMOISELLE BRANDÈS (Nos 8331, 8331, 8445, 8445, 8446 8447, — *Toilettes de Soirées*)

VI

Photoplastographie Patin — Bois-Colombes

Plate 14. Paul Nadar, Vignettes of three actresses, from Paris Photographie, *no. 2, May 1891. This Woodburytype print shows no signs of aging characteristic of many photographs from this time period compared with the albumen print at plate 13 published the following month in the same journal.*

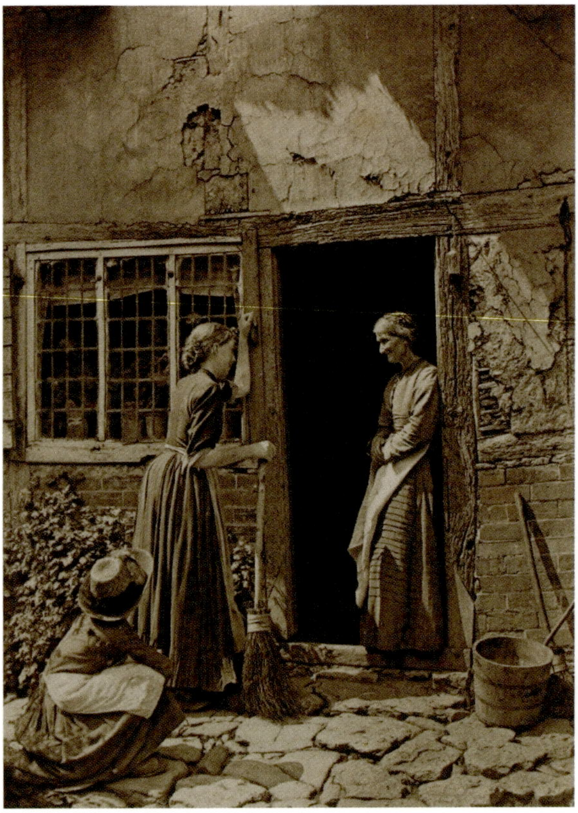

Plate 15a. Photogravure, Village News, *photograph by J. Gale, printed by Dawson's Photography, published in* Photography: Its History, Processes, Apparatus and Materials, *1892 by A. Brothers. Notice the softness of the details compared with the Woodburytype at plate 16 published in the same book.*

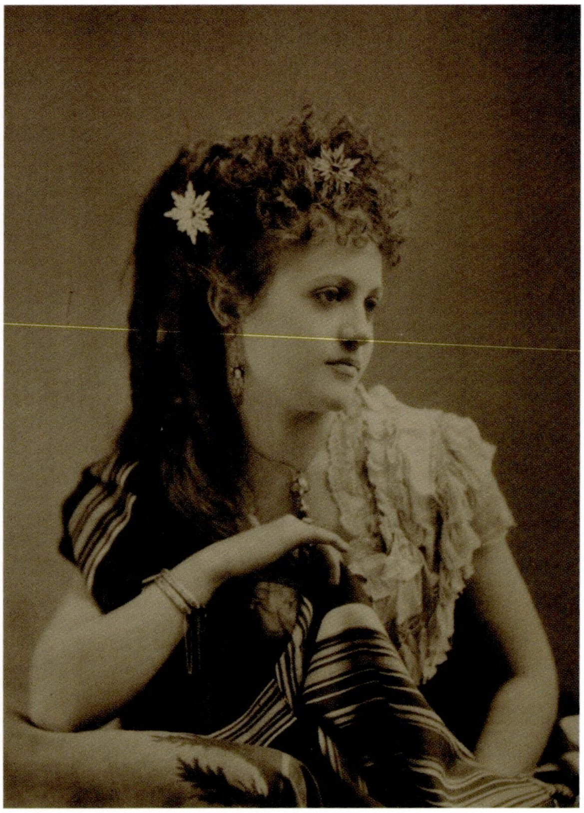

Plate 15b. Collotype portrait printed by B.J. Edwards & Co.'s Photo-Tint process, published in History and Handbook of Photography, *1876 by Gaston Tissandier. Notice the softness of the details compared with the Woodburytype at plate 17.*

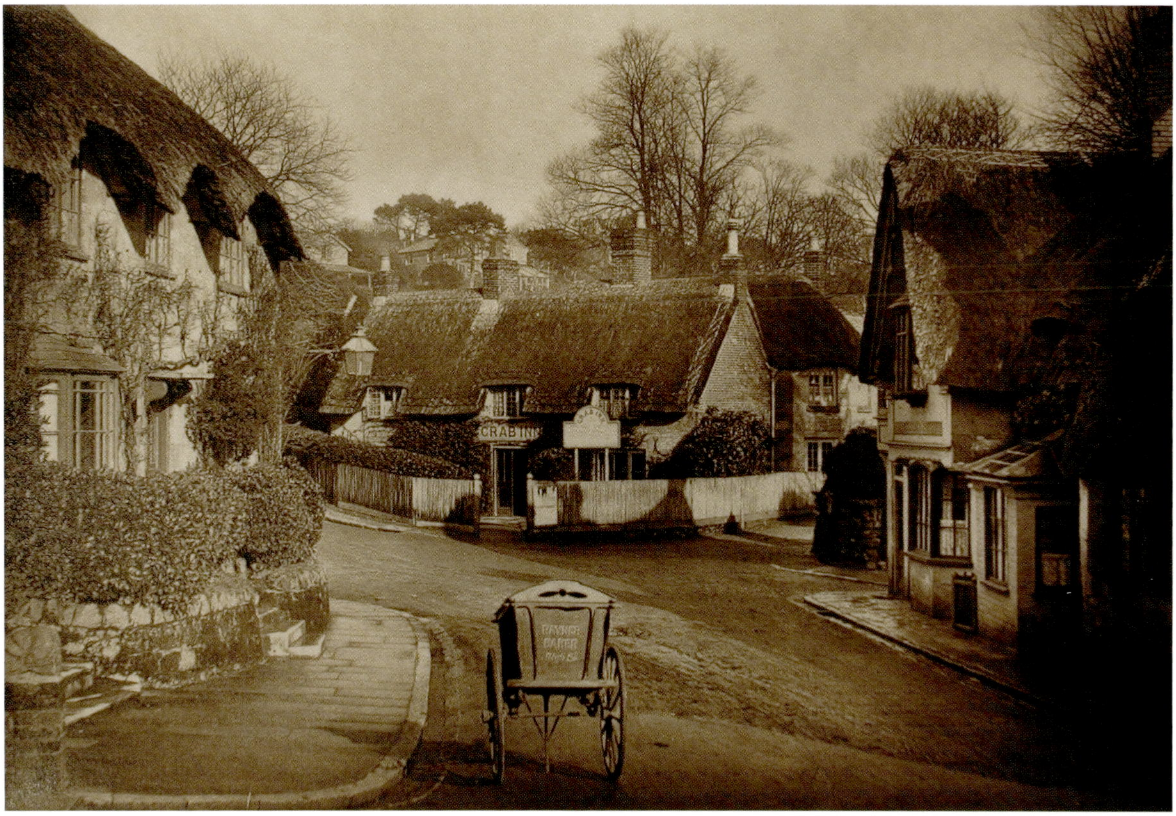

Plate 16. Woodburytype, Crab Inn, Shanklin, Isle of Wight, *photograph by Charles J. Brennan, published in* Photography: Its History, Processes, Apparatus and Materials, *1892 by A. Brothers.*

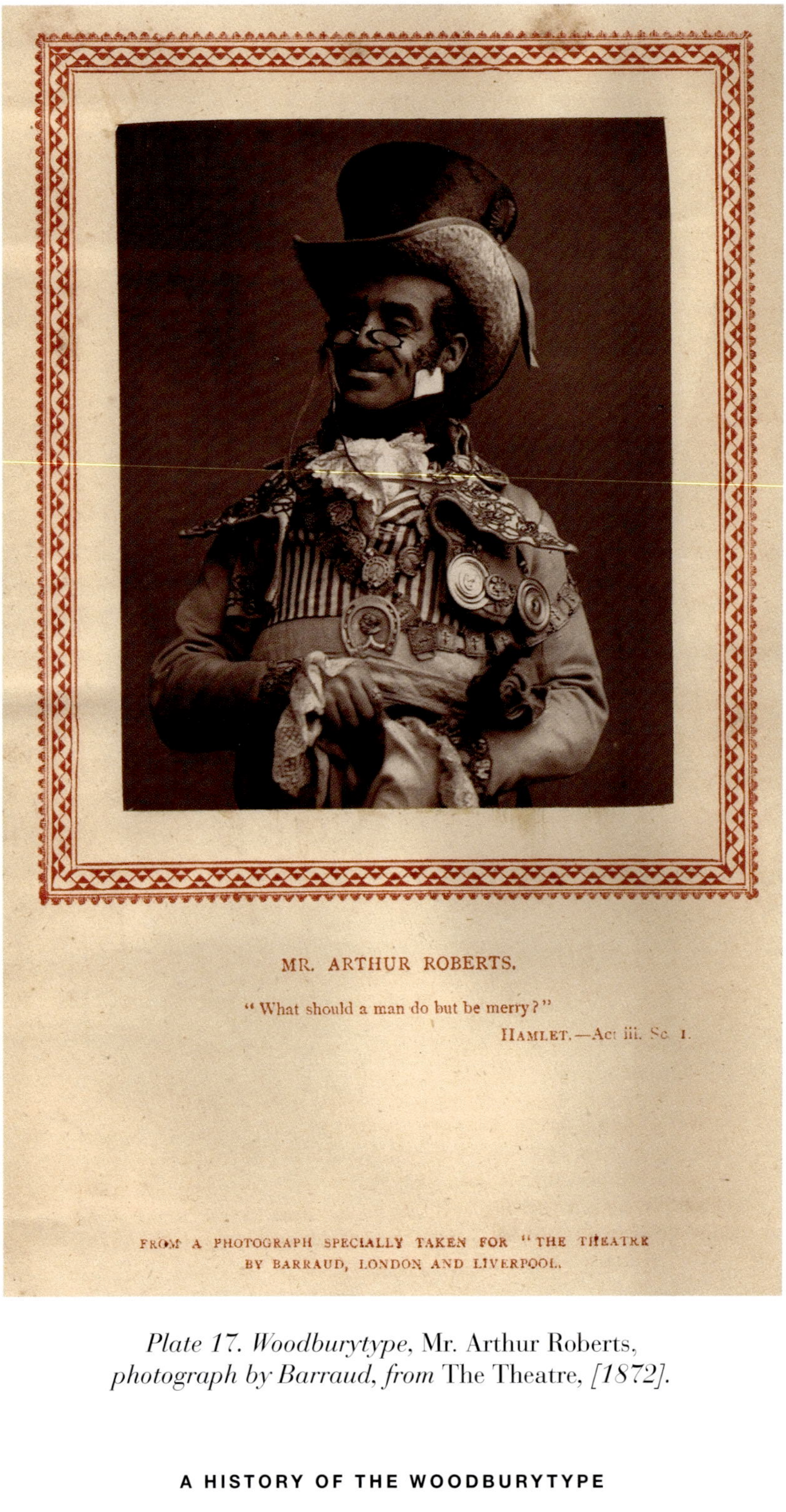

Plate 17. Woodburytype, Mr. Arthur Roberts, *photograph by Barraud, from* The Theatre, *[1872].*

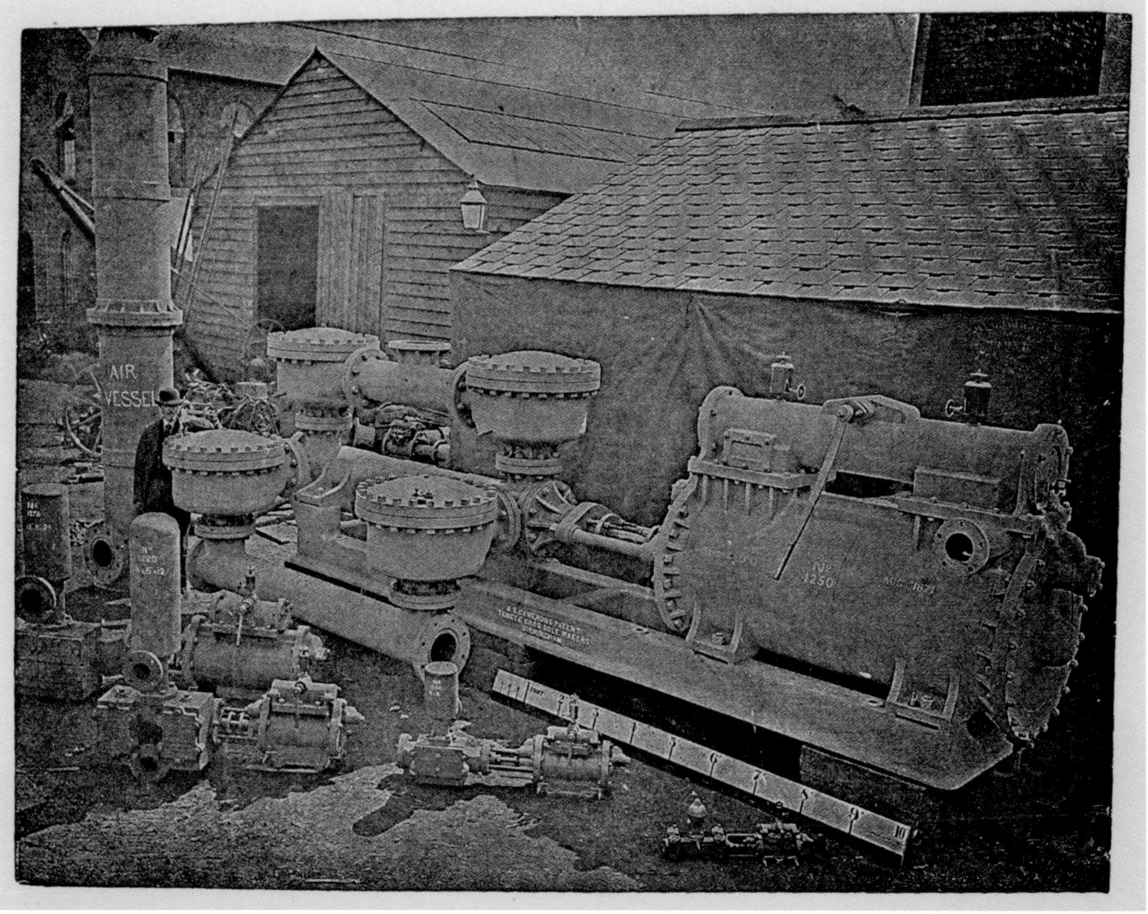

Plate 18. Woodbury's screen process printed in ordinary printer's ink, photograph of the hydraulic apparatus used to run the Woodburytype press, manufactured by Tangye Brothers. The block was made [1873], the print was published in the Photographic News, *October 26, 1883.*

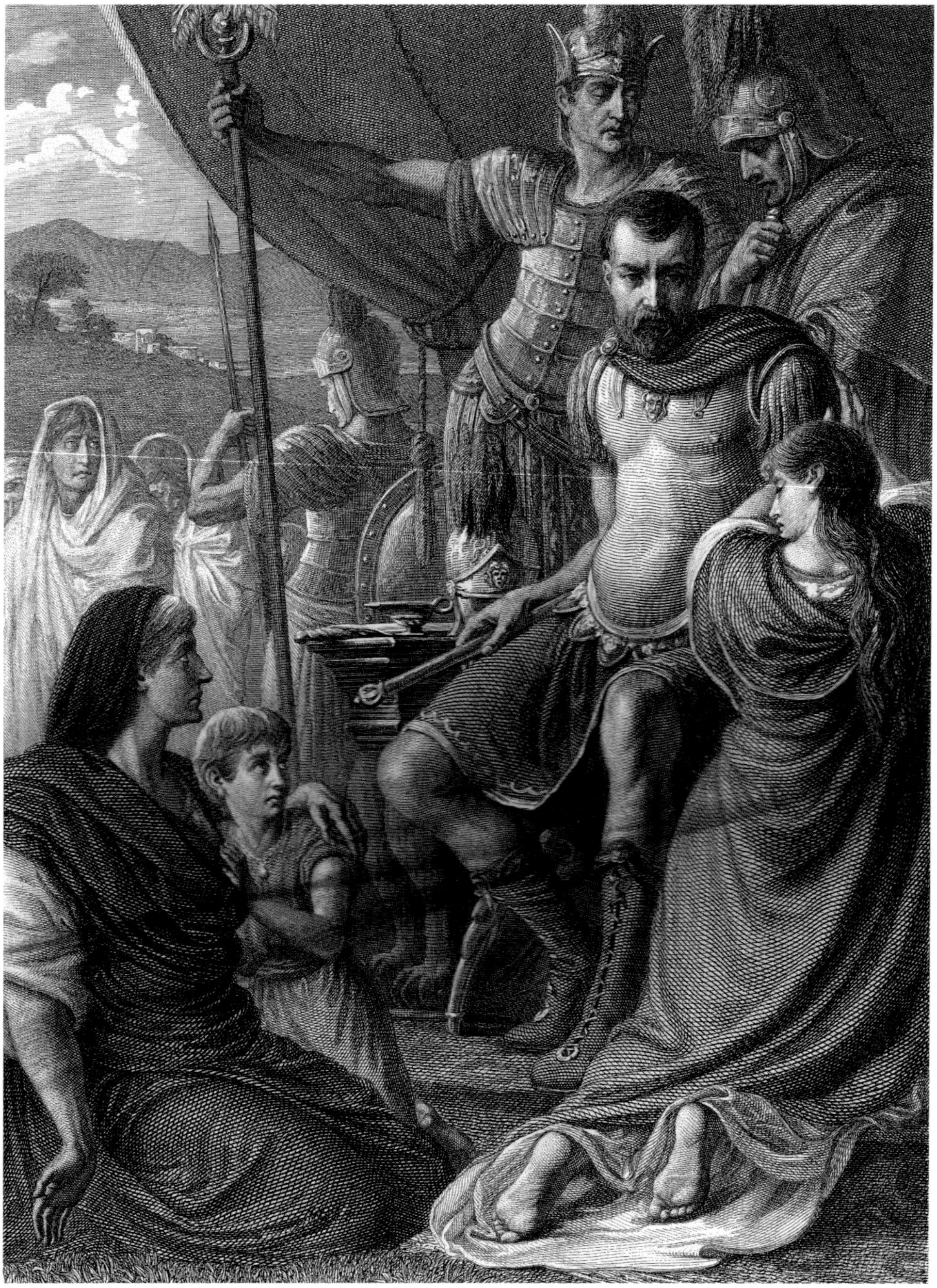

Plate 19. Metal engraving, Volumnia and Coriolanus, *by H.A. Bone, engraved by Job. Bauer.*

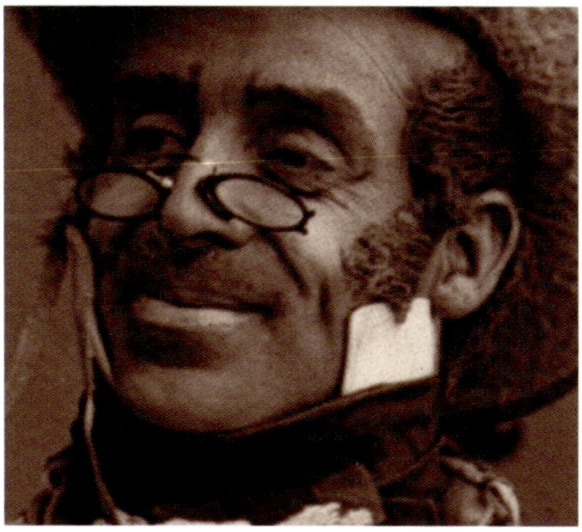

Plate 20a. Detail of plate 17, enlarged about four times. Notice the subtleties of gradation in the middle values compared with plates 20b and 20c.

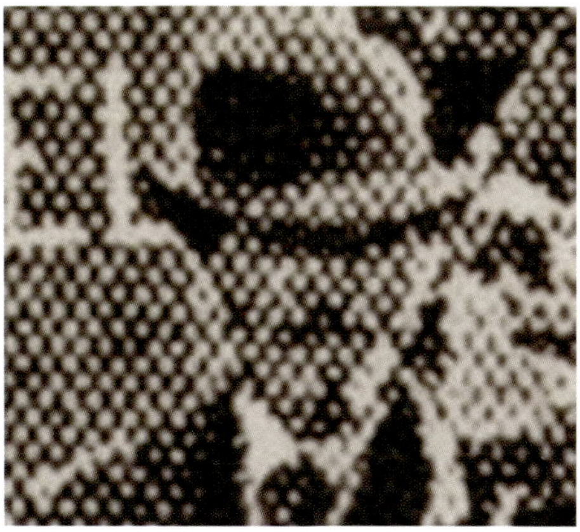

Plate 20b. Detail of plate 18, enlarged about four times. Notice the distinct squares of black ink and white paper-base used to produce the illusion of gradation in the middle values compared with plate 20a.

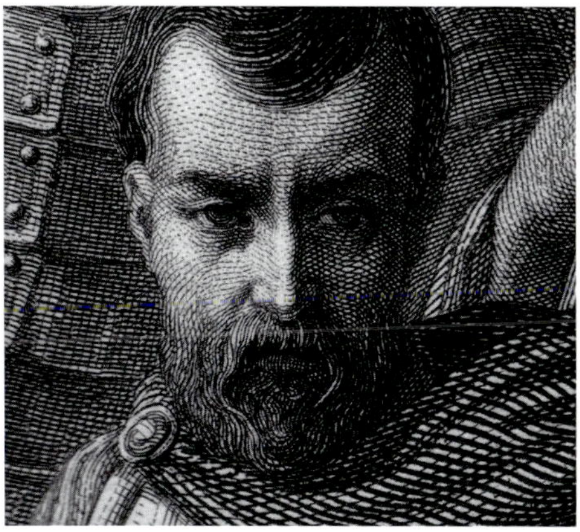

Plate 20c. Detail of plate 19, enlarged about four times. Notice the distinct lines of black ink and white paper-base used to produce the illusion of gradation in the middle values compared with plate 20a.

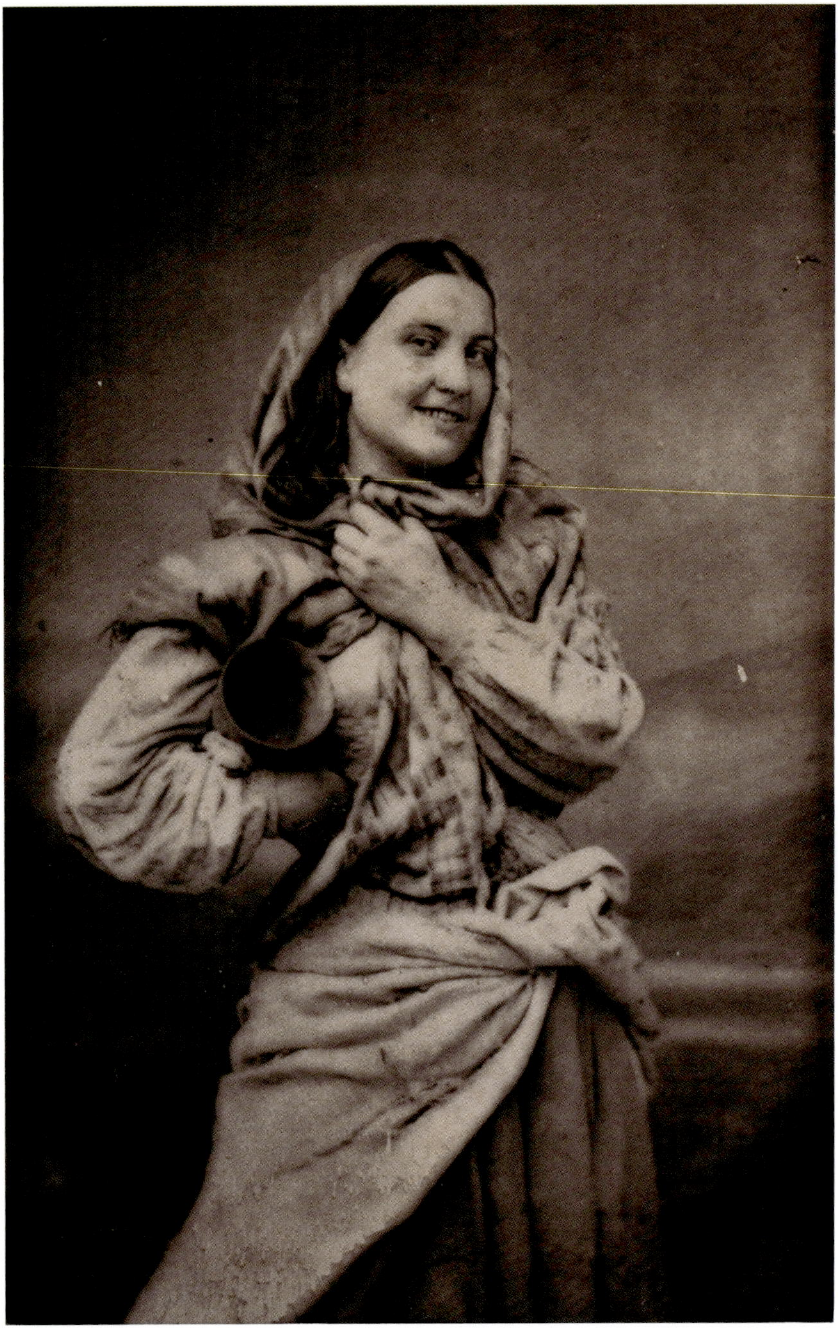

Plate 21. H.P. Robinson, Mountain Dew Girl: Kilarney, *first published Woodburytype print, published in the* Photographic News, *January 1866.*

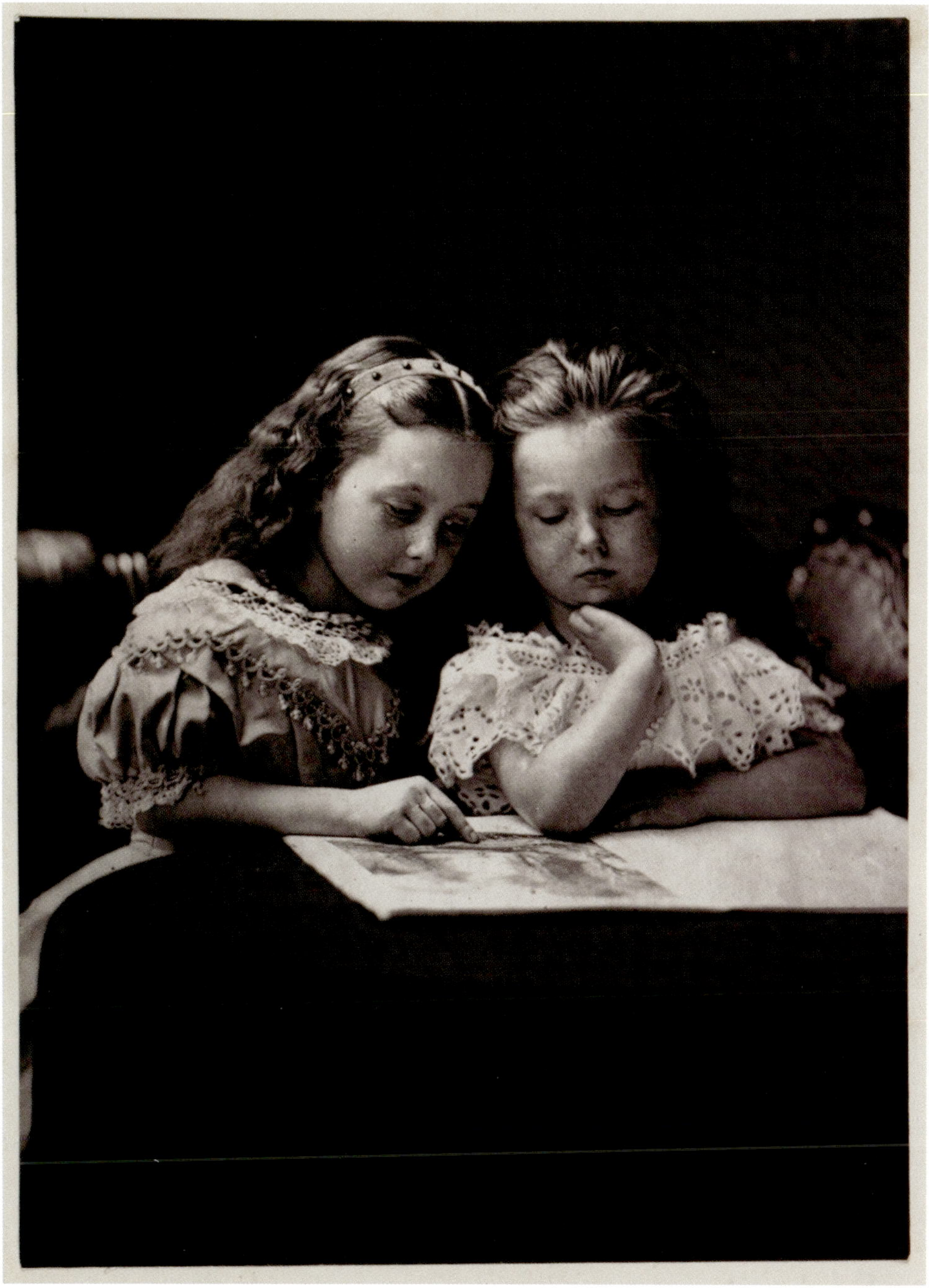

Plate 22. H. P. Robinson, Maud and Ethel Robinson Reading, *Woodburytype frontispiece to Robinson's* Pictorial Effect in Photography, *1869.*

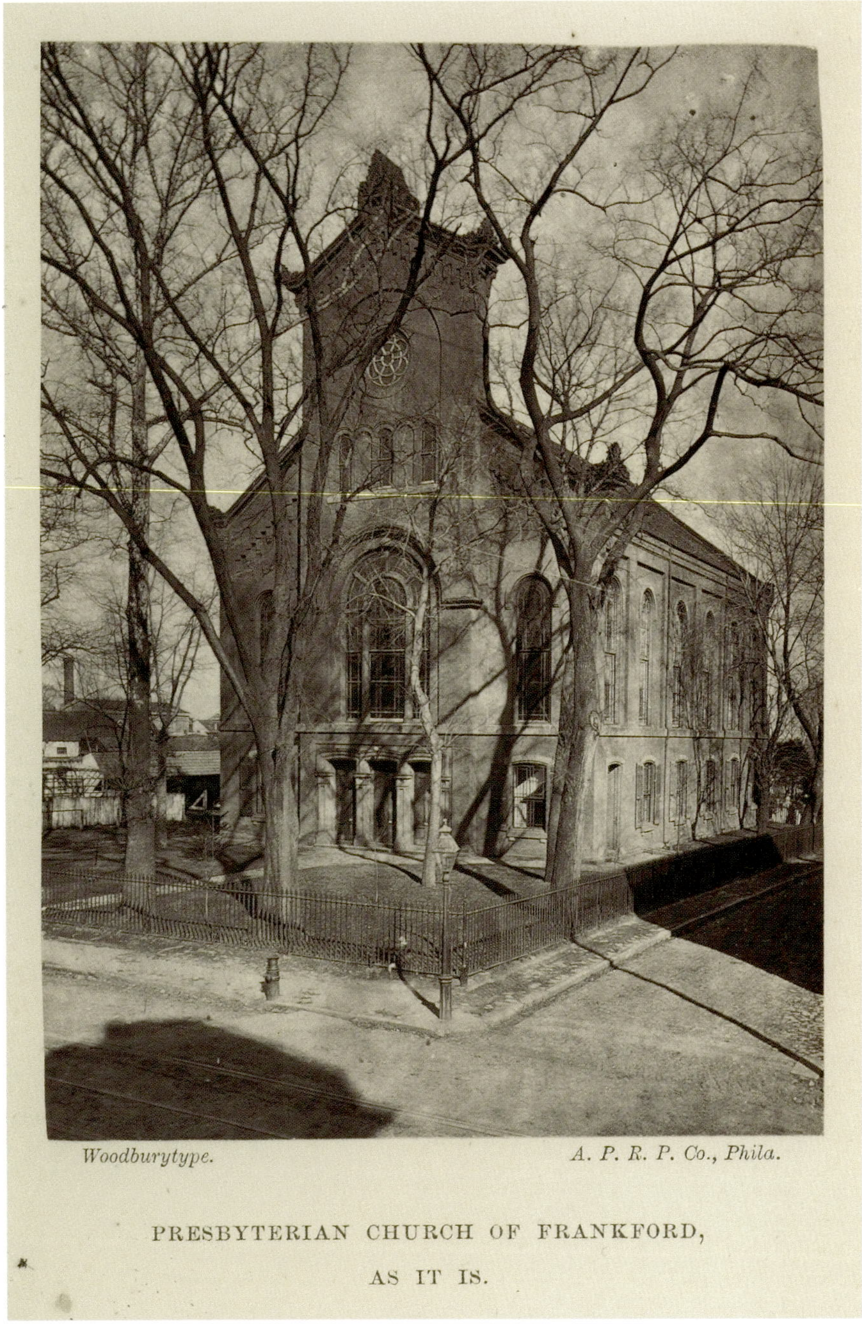
Woodburytype. *A. P. R. P. Co., Phila.*

PRESBYTERIAN CHURCH OF FRANKFORD,

AS IT IS.

Plate 23. Presbyterian Church of Frankford, one of two Woodburytype frontispieces from One Hundred Years of the Presbyterian Church in Frankord, *1872, by Rev. Thomas Murphy. Printed by John Carbutt's American Photo-Relief Printing Company.*

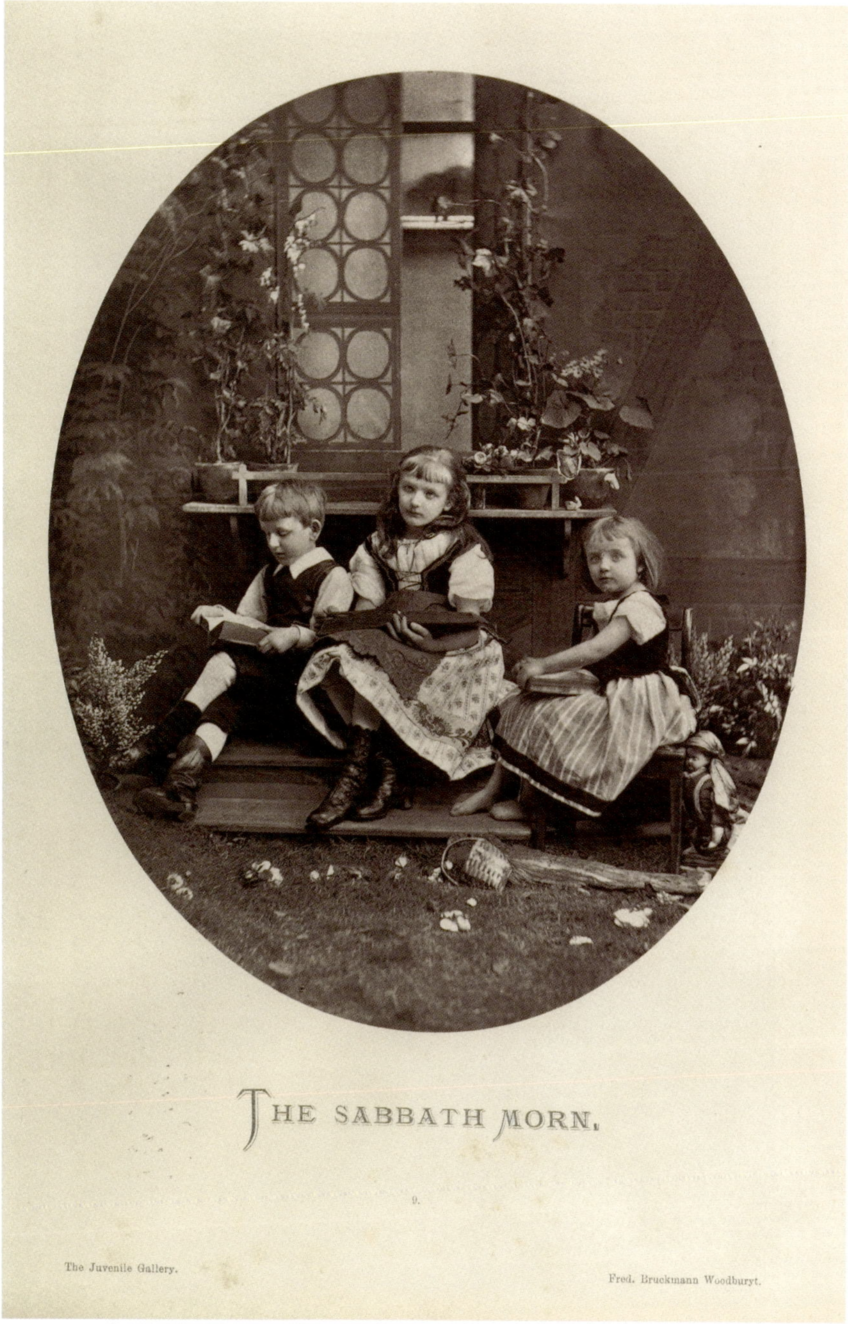

Plate 24. The Sabbath Mourn, *Woodburytype illustration from* The Juvenile Gallery, *[1889].*

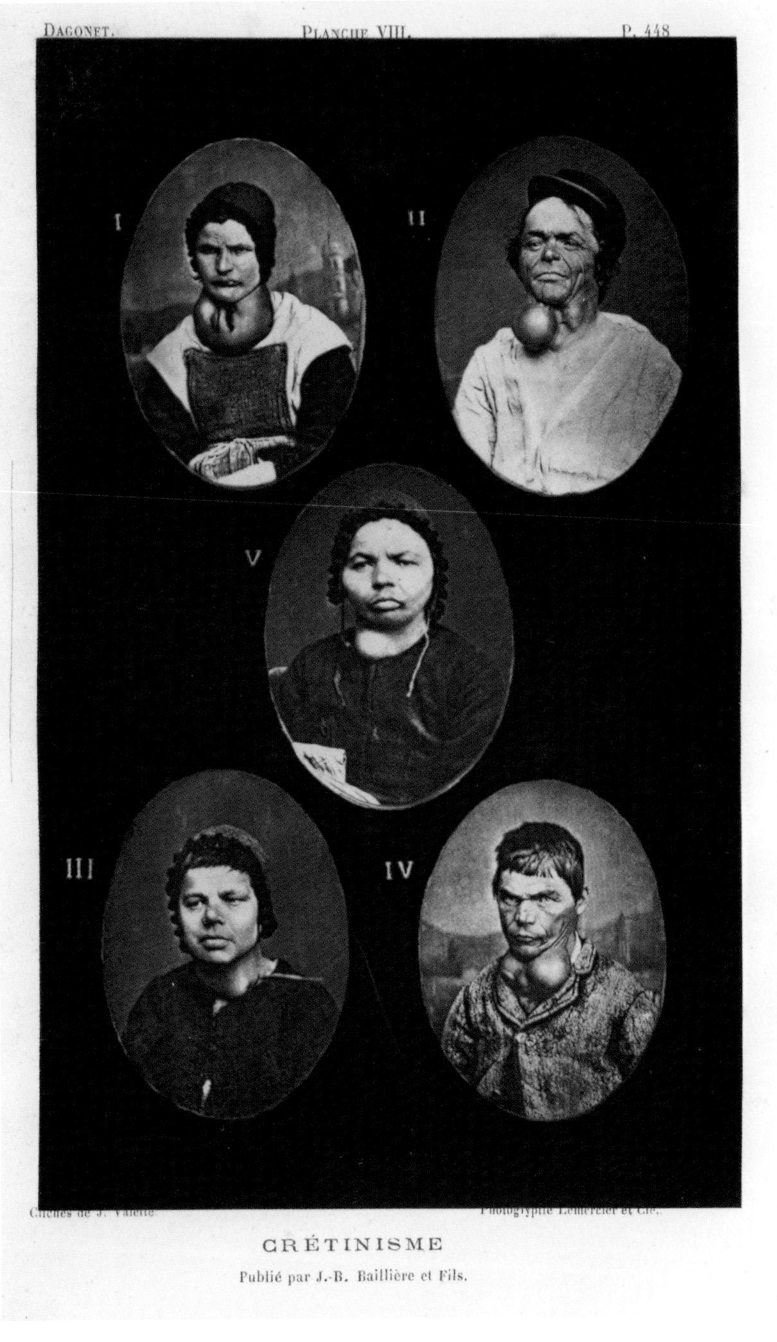

Plate 25. Cretinisme, *photograph by J. Valette, printed by Lemercier, Woodburytype illustration from* Nouveau Traite Elementaire at Pratique des Maladies Mentales, *1876, by H. Dagonet.*

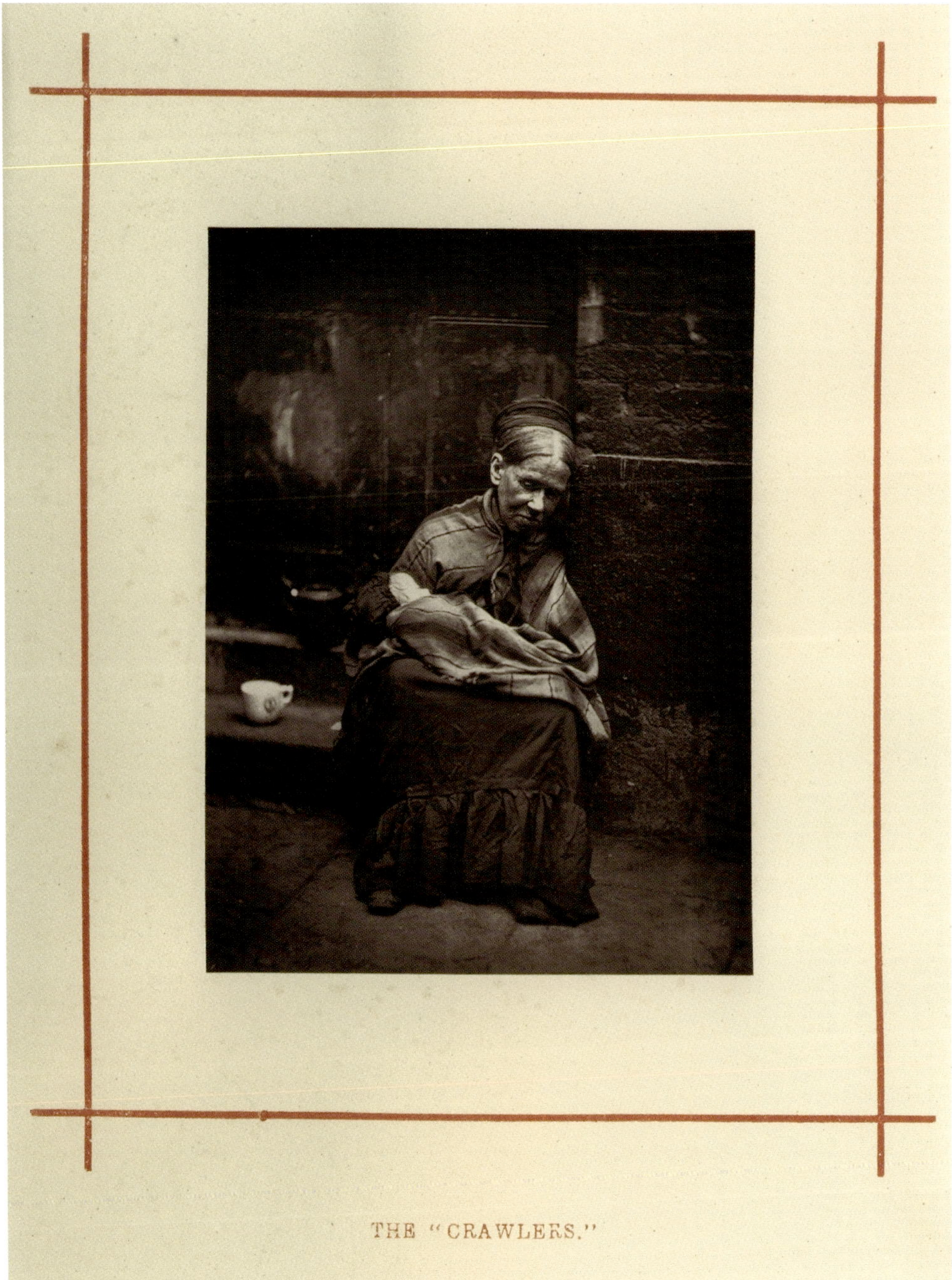

Plate 26. John Thomson, The Crawlers, *Woodburytype from* Street Life in London, *1877.*

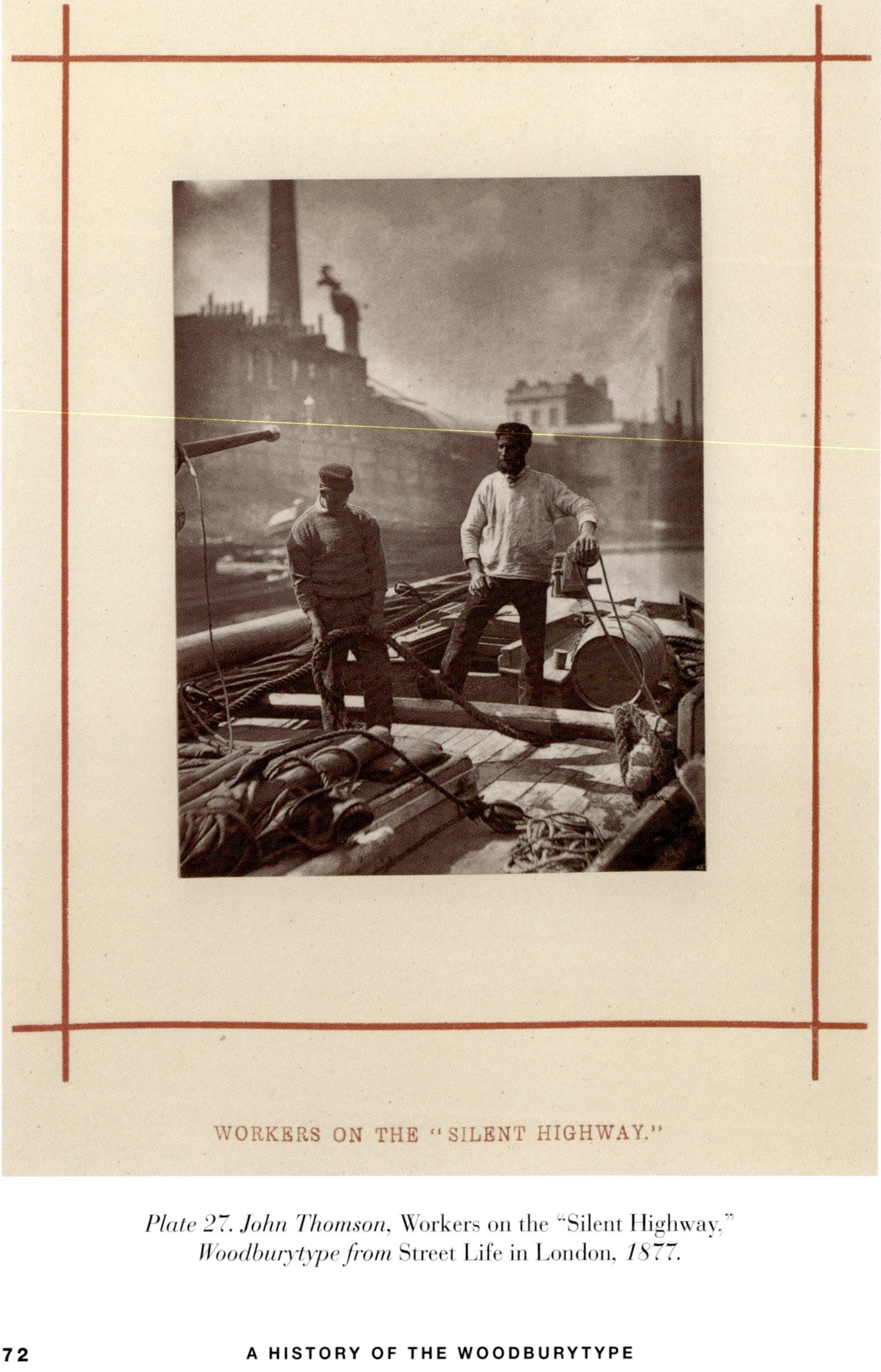

Plate 27. John Thomson, Workers on the "Silent Highway," *Woodburytype from* Street Life in London, *1877.*

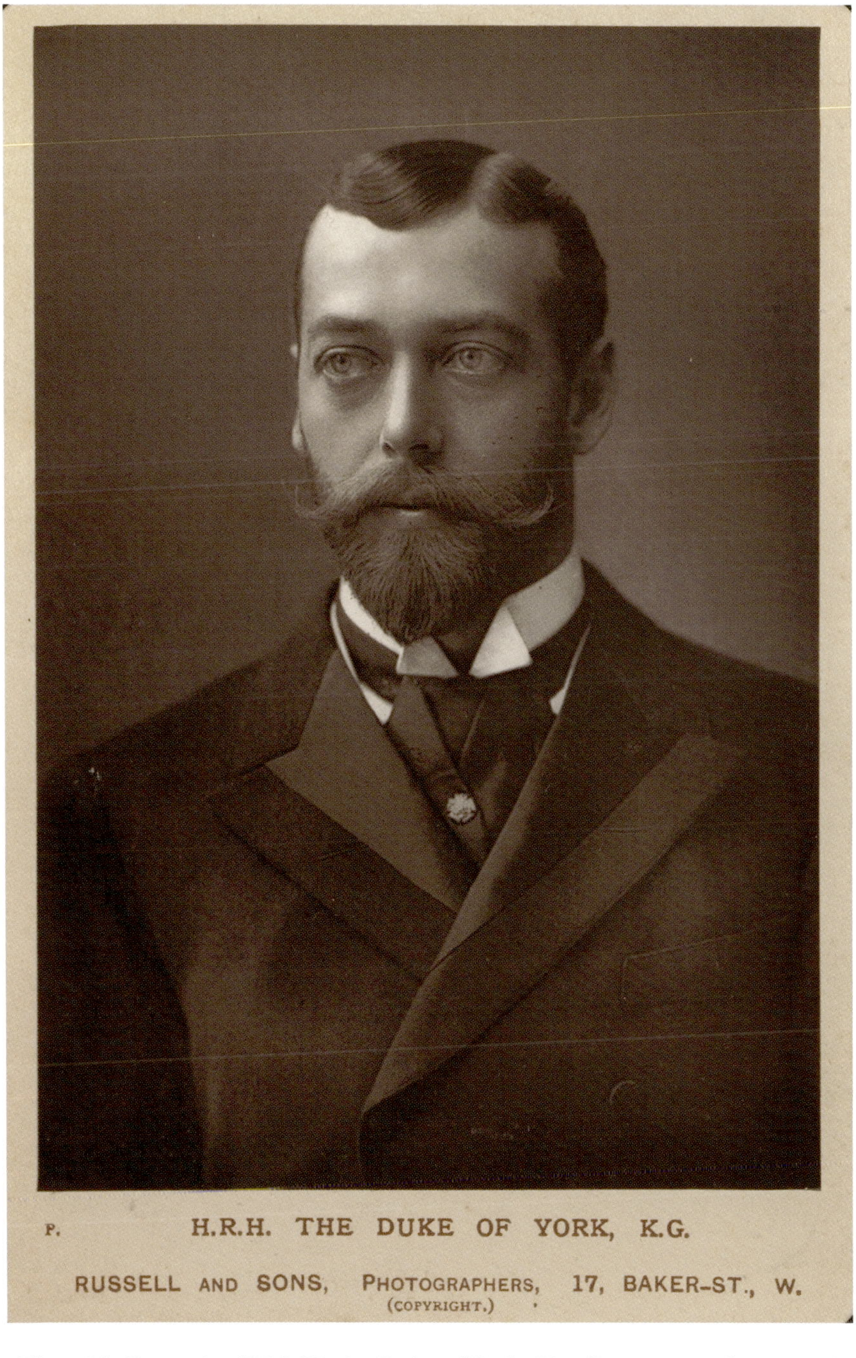

Plate 28. Portrait of H.R.H. the Duke of York, Woodburytype cabinet card, photograph by James Russell and Sons, [1890].

SARAH BERNHARDT

Mademoiselle SARAH BERNARDT a été bien inspirée le jour où elle a ajouté à son chiffre enlacé une banderolle flottante sur laquelle on lit ces mots : « *Quand même.* » Elle ne pouvait pas résumer sa vie et sa personnalité d'une manière plus complète.

Dans ce défi altier, jeté à tous les obstacles, on reconnaît la volonté énergique de l'artiste qui donne tête baissée dans les plus grandes aventures dramatiques, sa persévérance dans la voie choisie, son ambition d'arriver au but malgré tout, d'étreindre le rêve si haut, si loin qu'il puisse être. Légende orgueilleuse, dira-t-on. Parbleu ! mais il y a dans ce « *Quand même!* » plus et mieux que de l'orgueil. Il y a, la noble fierté de soi-même, le dédain de l'impossible et le mépris des sots et des méchants.

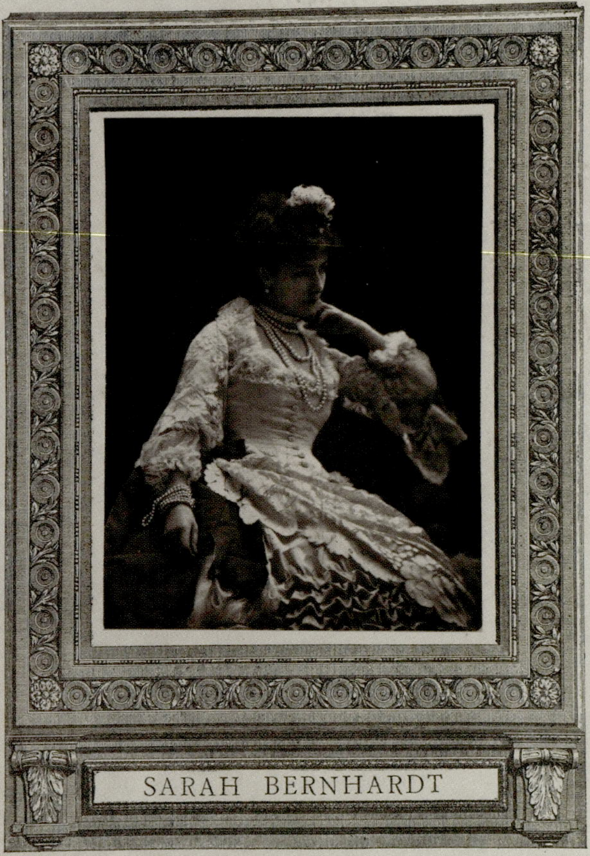

Quand on passe en revue les phases diverses de la carrière artistique de mademoiselle Sarah Bernhardt, on voit combien elle a toujours été fidèle à sa devise. Jamais légende accompagnant des armoiries séculaires n'a été suivie plus scrupuleusement que la sienne.

Aussi loin que l'on remonte dans le passé de l'artiste, qu'on la prenne à ses premiers débuts au théâtre, ou même à l'époque où elle n'était encore qu'une petite pensionnaire, on retrouve des souvenirs, des faits, des mots qui sont du ressort du fameux « quand même ! »

M. Francisque Sarcey a publié dernièrement une biographie très-intéressante de mademoiselle Sarah Bernhardt.

Après avoir parlé de l'éducation que la charmante sociétaire du Théâtre-Français reçut à Versailles, au couvent de Grandchamp, l'un des établissements les plus aristocratiques de France, le critique du *Temps* ajoute ceci :

« Il lui a été donné de voir plus tard les notes qui lui avaient été appliquées au couvent. L'une d'elles exprimait, dans le langage de l'endroit, qu'elle n'était point née pour les conditions ordinaires de la vie, qu'elle deviendrait fatalement ou l'une des lumières les plus vives de la piété, ou l'une des plus terribles pierres de scandale que la religion ait jamais anathématisées. »

Quand elle sortit du couvent, chargée de couronnes, et qu'il lui fut pour la première fois permis de formuler ses souhaits pour l'avenir :

« Je veux être religieuse, » s'écria-t-elle d'un ton de foi ardente. Et comme on se regardait avec un air de stupéfaction : « A moins, ajouta-t-elle, que je ne sois actrice. »

Ce trait de caractère me dispense d'insister sur les entraînements et les aspirations de l'artiste. Une enfant qui parle ainsi, qui se prononce si franchement pour l'un ou l'autre des points extrêmes, la première fois qu'elle regarde la vie et l'avenir, ne saurait suivre une voie banale. Elle s'annonce en conquérante. Elle doit arriver quand même.

Je continue à citer M. Francisque Sarcey :

« C'est alors que l'on songea à la mettre au Conservatoire. Vous savez qu'il n'est pas possible d'entrer dans cette école de l'État sans passer un examen préliminaire : chaque candidat choisit un morceau à sa convenance et le récite devant un jury, qui l'accepte ou le repousse. La mère de mademoiselle Sarah Bernhardt était peu familière avec les exigences de la maison ; sa fille savait la fable des *Deux Pigeons*, de la Fontaine. On s'imagina

GALERIE CONTEMPORAINE, 126, boulevard de Magenta, Paris — Cliché TOURTIN, 32, rue Louis-le-Grand.

Plate 29. Tourtin, Portrait of Sarah Bernhardt, Woodburytype from Galerie Contemporaine, *[1880].*

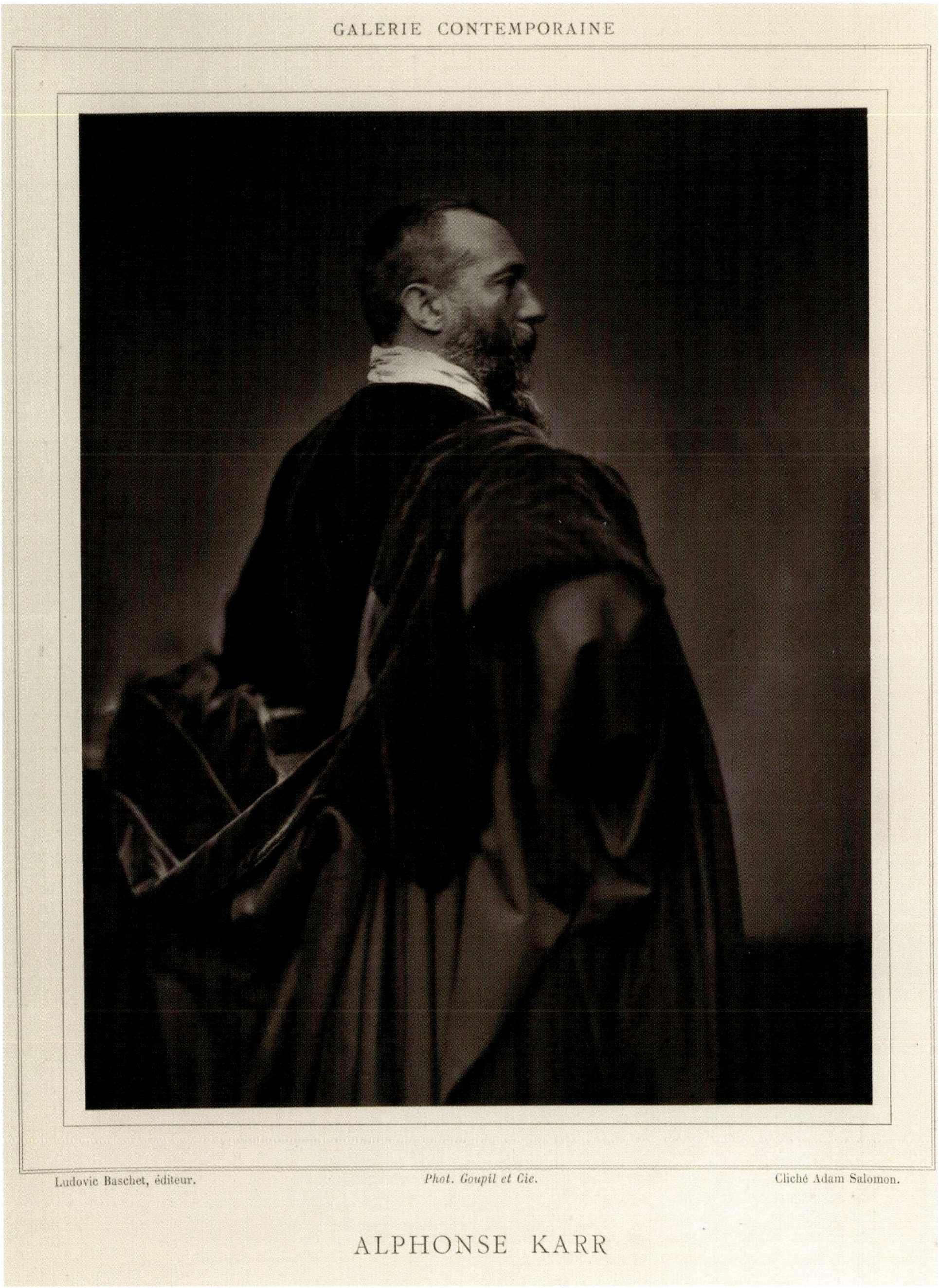

Plate 30. Adam Salomon, Portrait of Alphonse Karr, Woodburytype from Galerie Contemporaine, *[1880].*

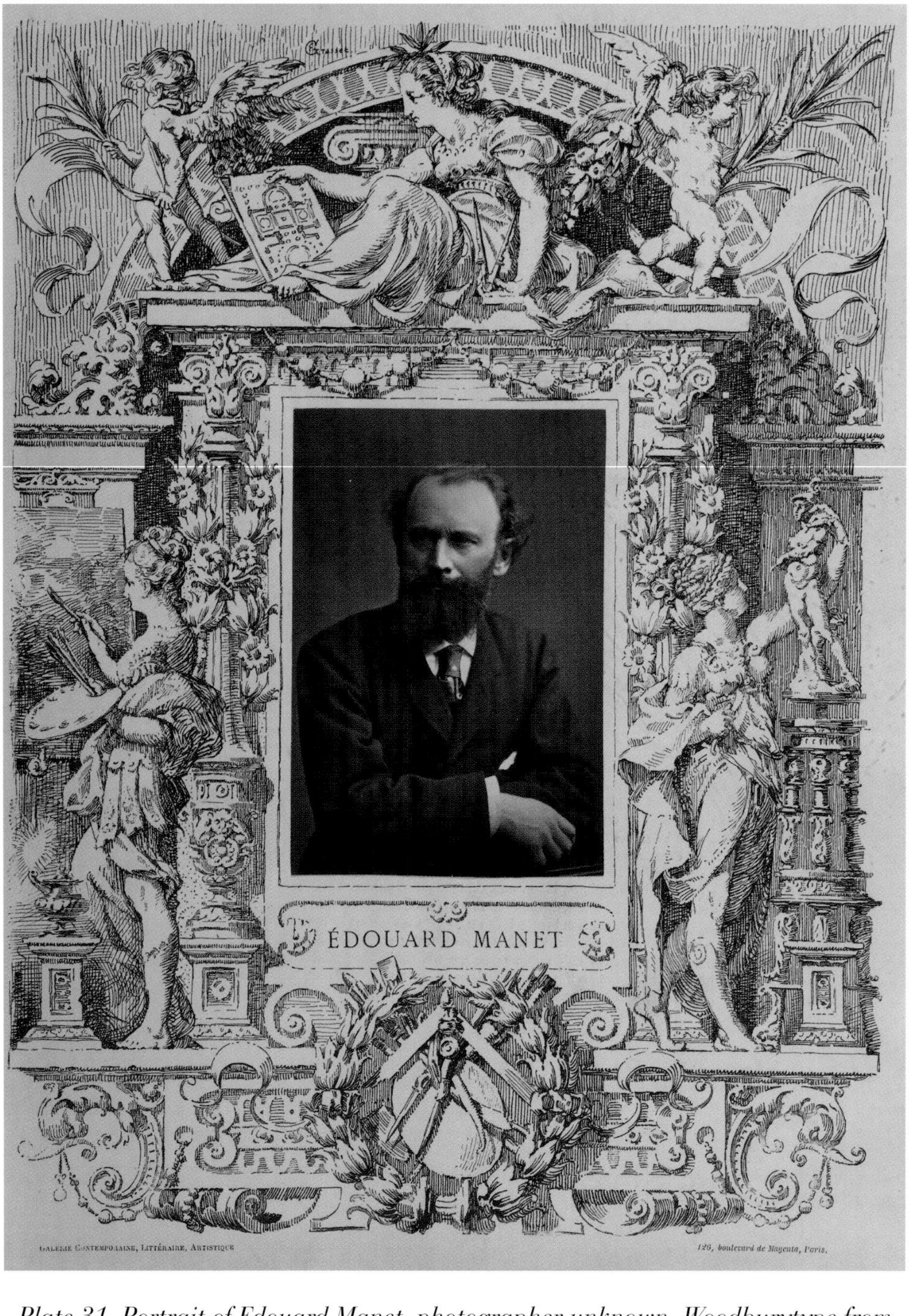

Plate 31. Portrait of Edouard Manet, photographer unknown, Woodburytype from Galerie Contemporaine, *[1880].*

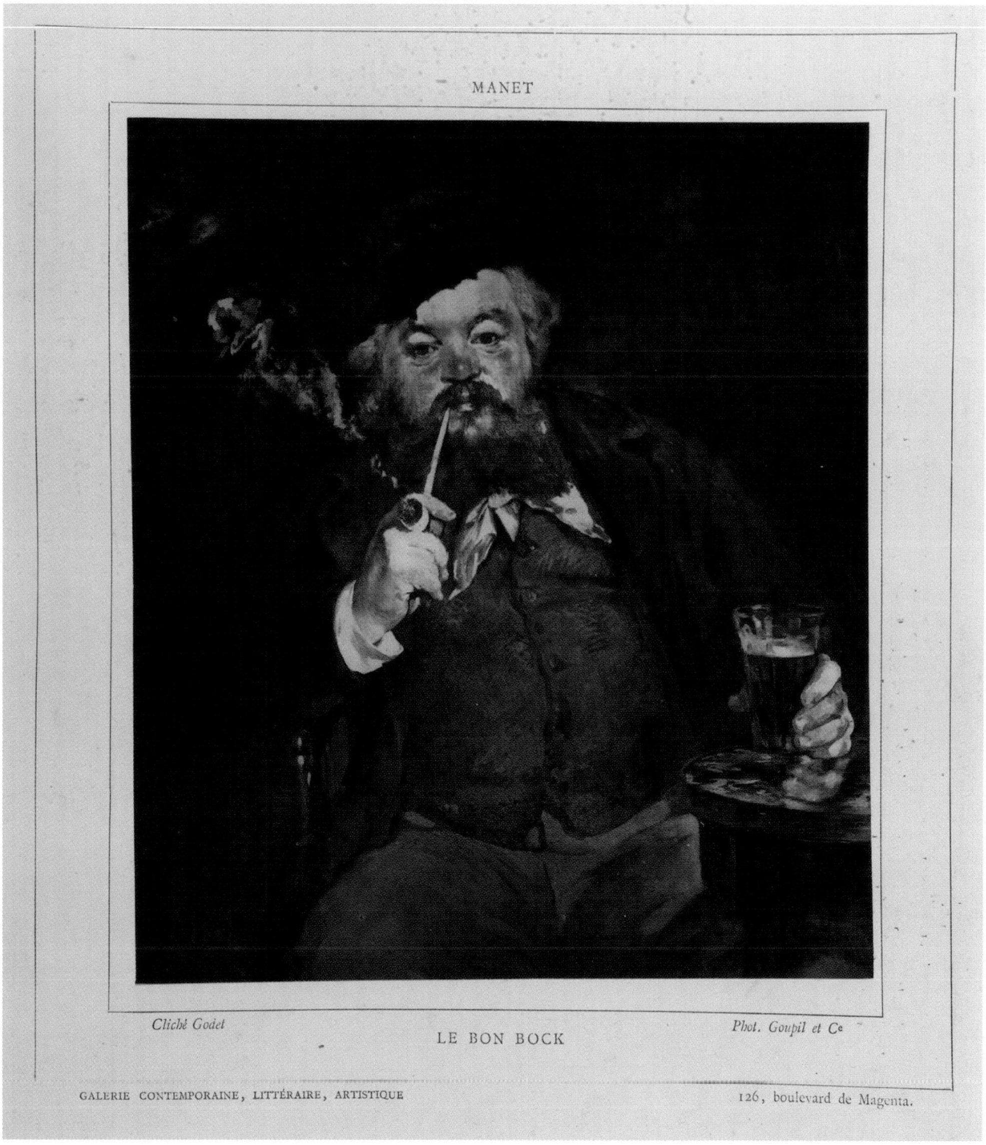

Plate 32. Woodburytype reproduction of Le Bon Bock, *by Manet, photographed by Godet, printed by Goupil, from* Galerie Contemporaine, *[1880].*

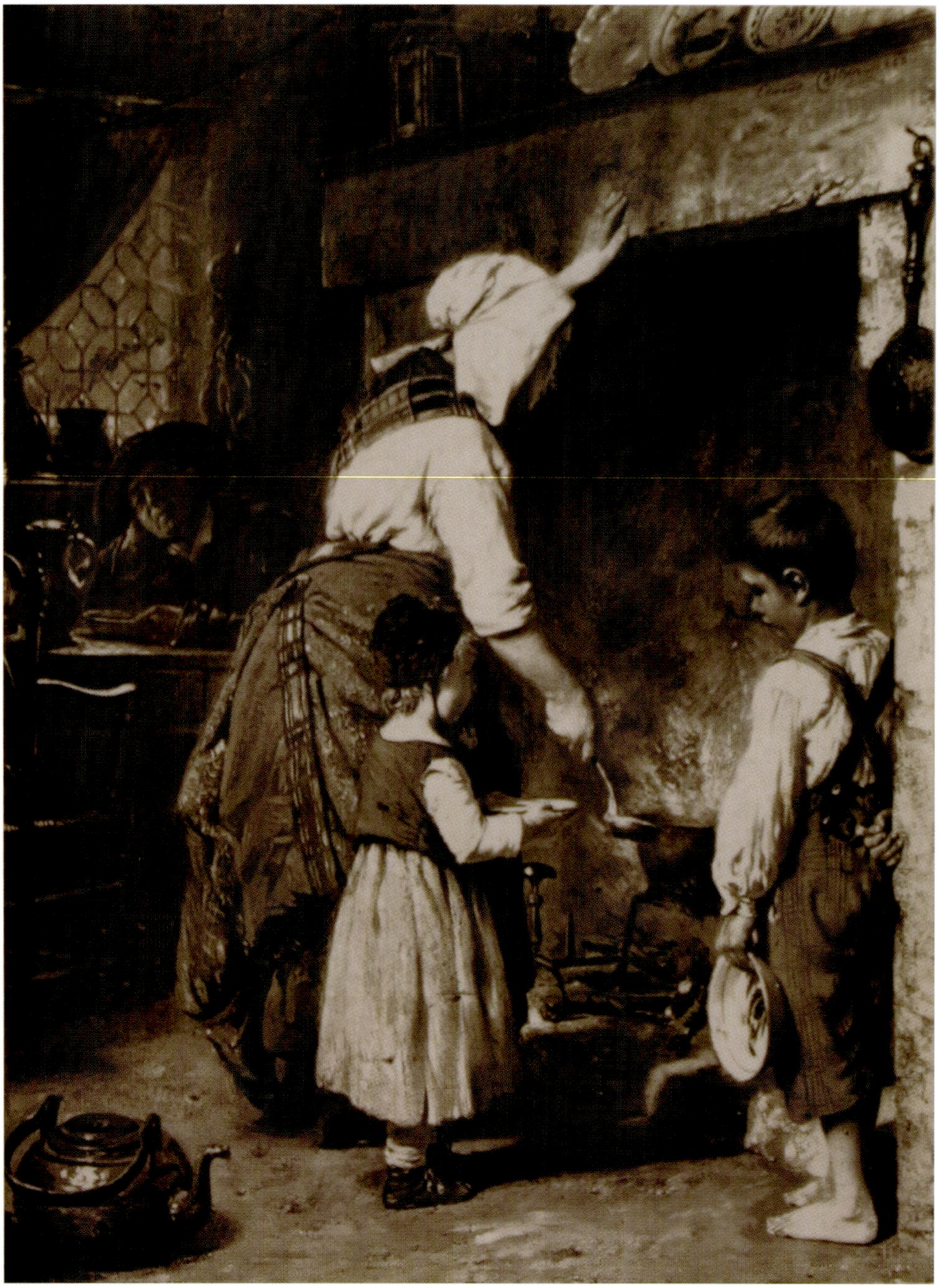

Plate 33. Woodburytype reproduction of a painting by Castan, printed by Lemercier, frontispiece to Les Merveilles de La Photographie, *1874 by Gaston Tissandier.*

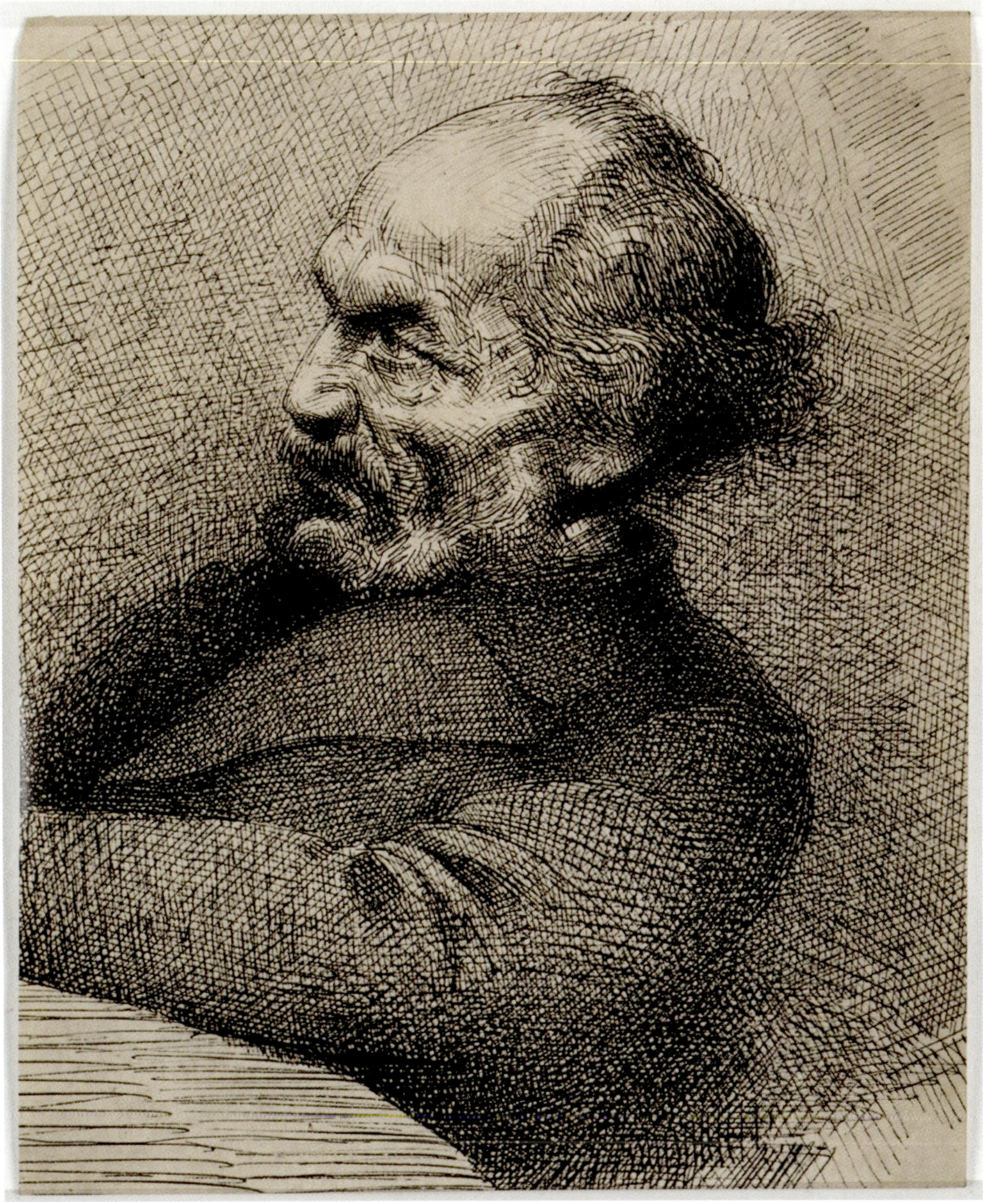

Plate 34. Woodburytype print of an ink drawing by an unknown maker.

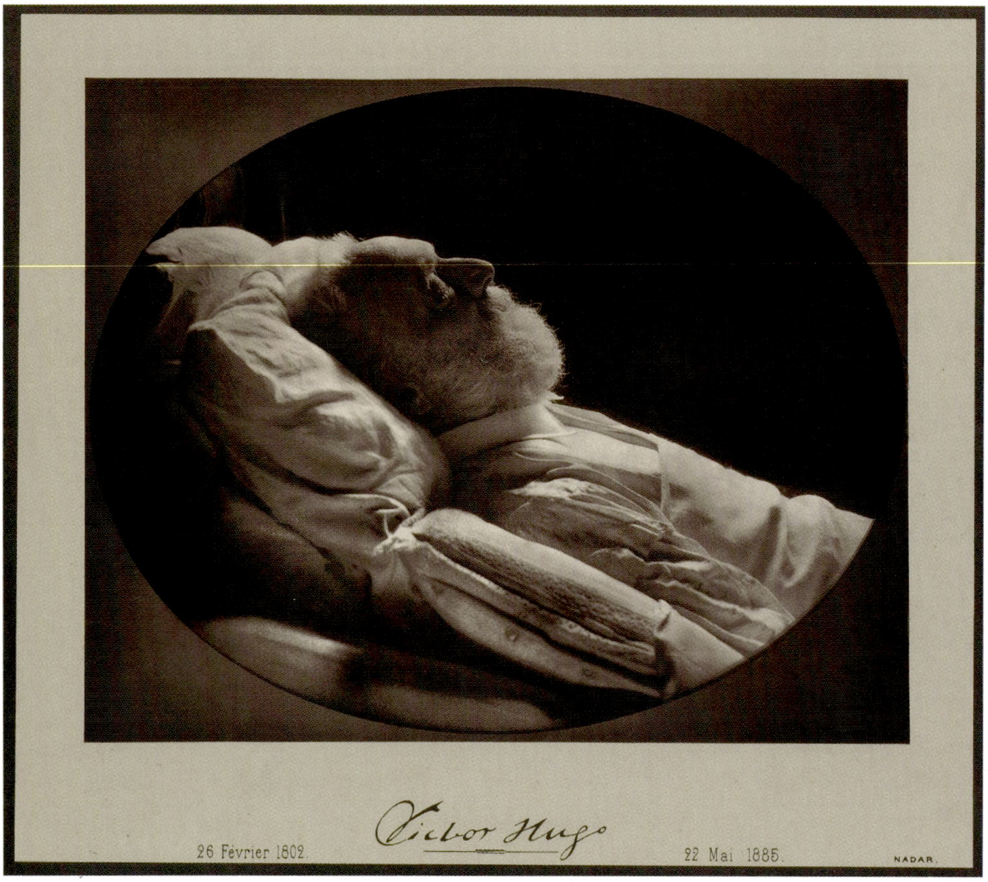

Plate 35. Nadar [Gaspard Felix Tournachon], Postmortem of Victor Hugo, *print ca. 1885.*

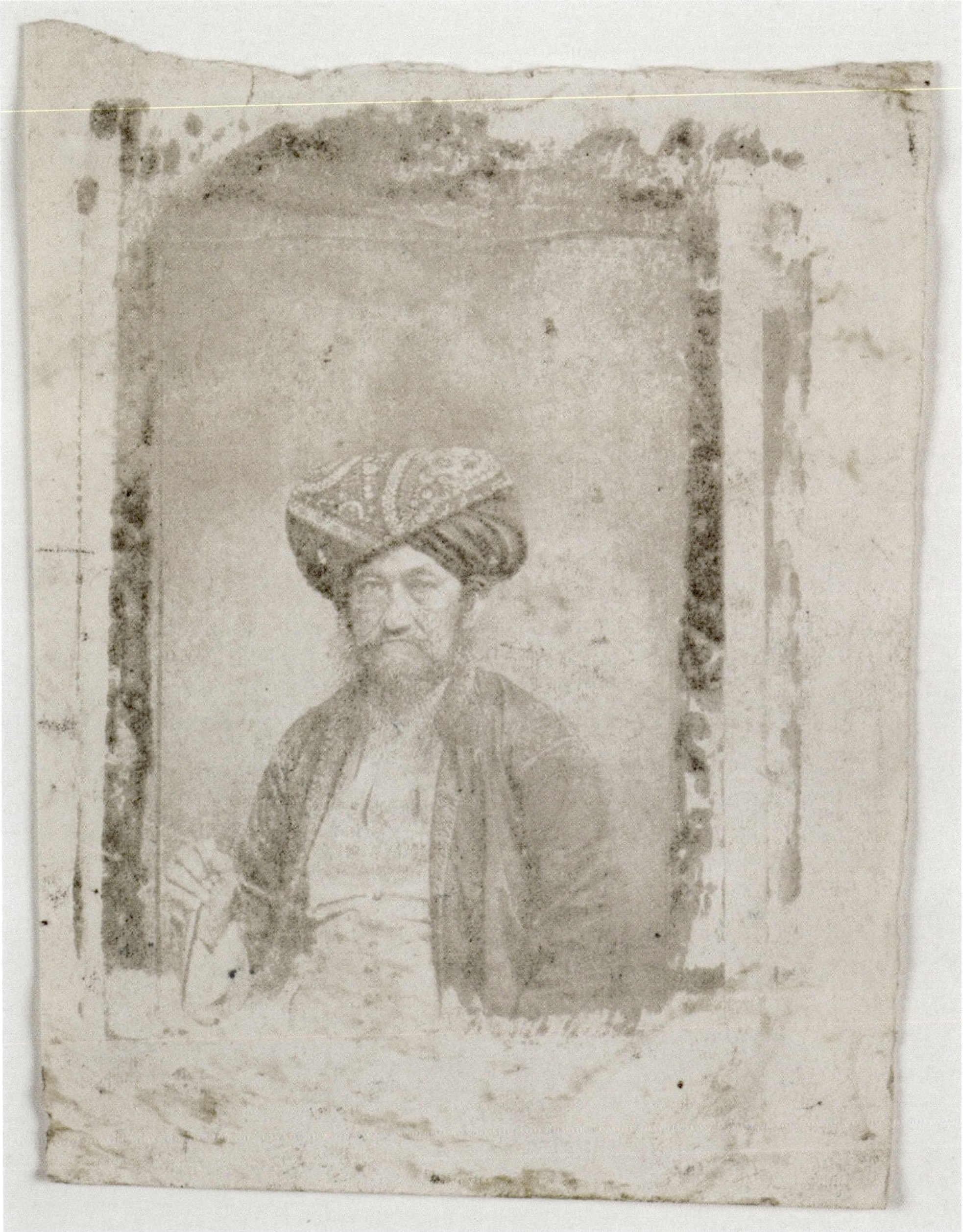

Plate 36. Unpublished very early experimental Woodburytype print on uncoated paper. The back is inscribed "Print from Woodbury's first Woodburytype mould." Notice the lack of density caused by the gelatin soaking into the paper-base.

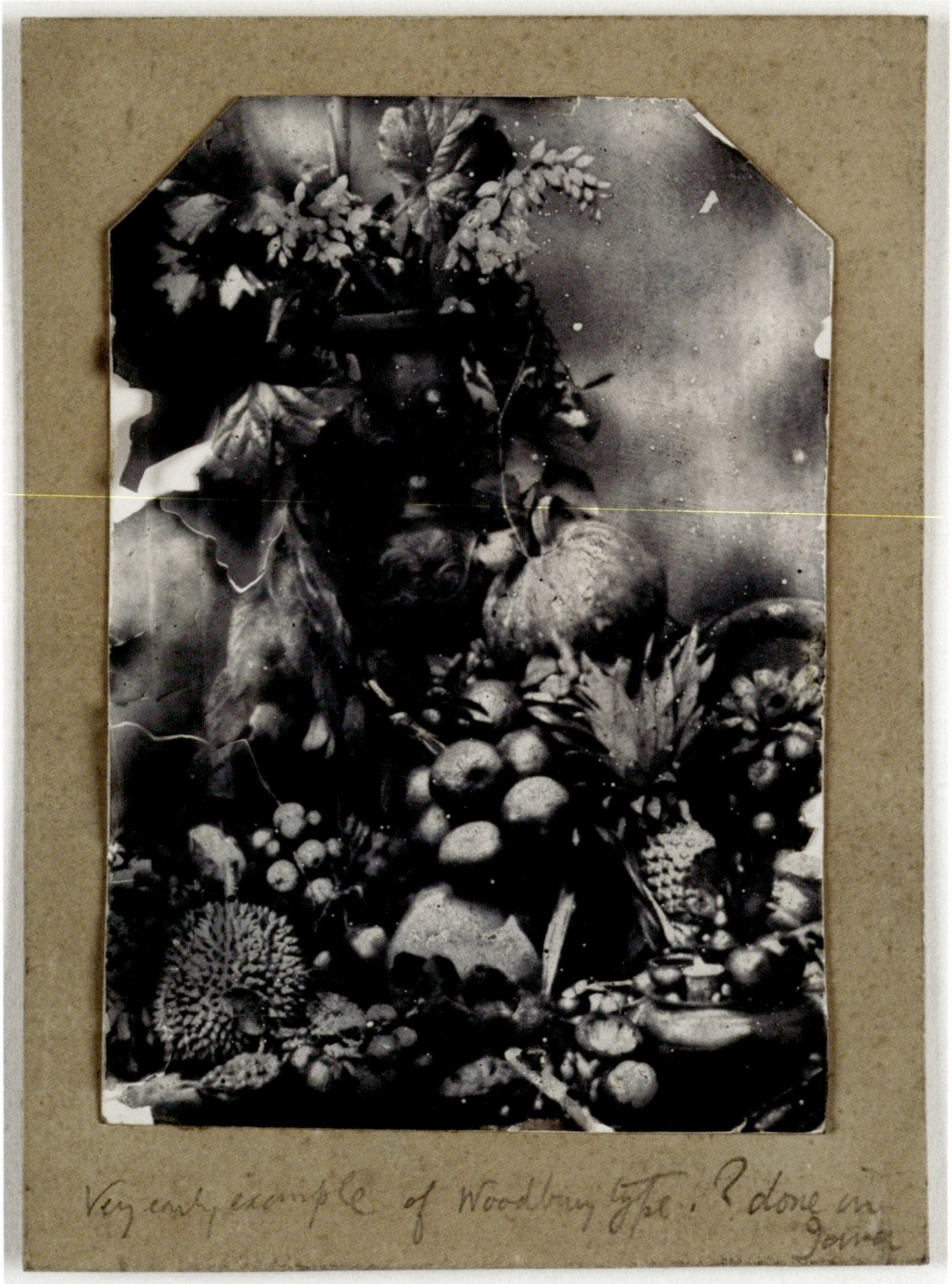

Plate 37. Unpublished very early experimental Woodburytype print of still life with fruit. Inscription by Woodbury reads "Very early example of Woodbury type. ? done in Java." Notice the cracking and peeling of the gelatin. This image is almost identical to plate 38, also notice the similarities to figure 1.

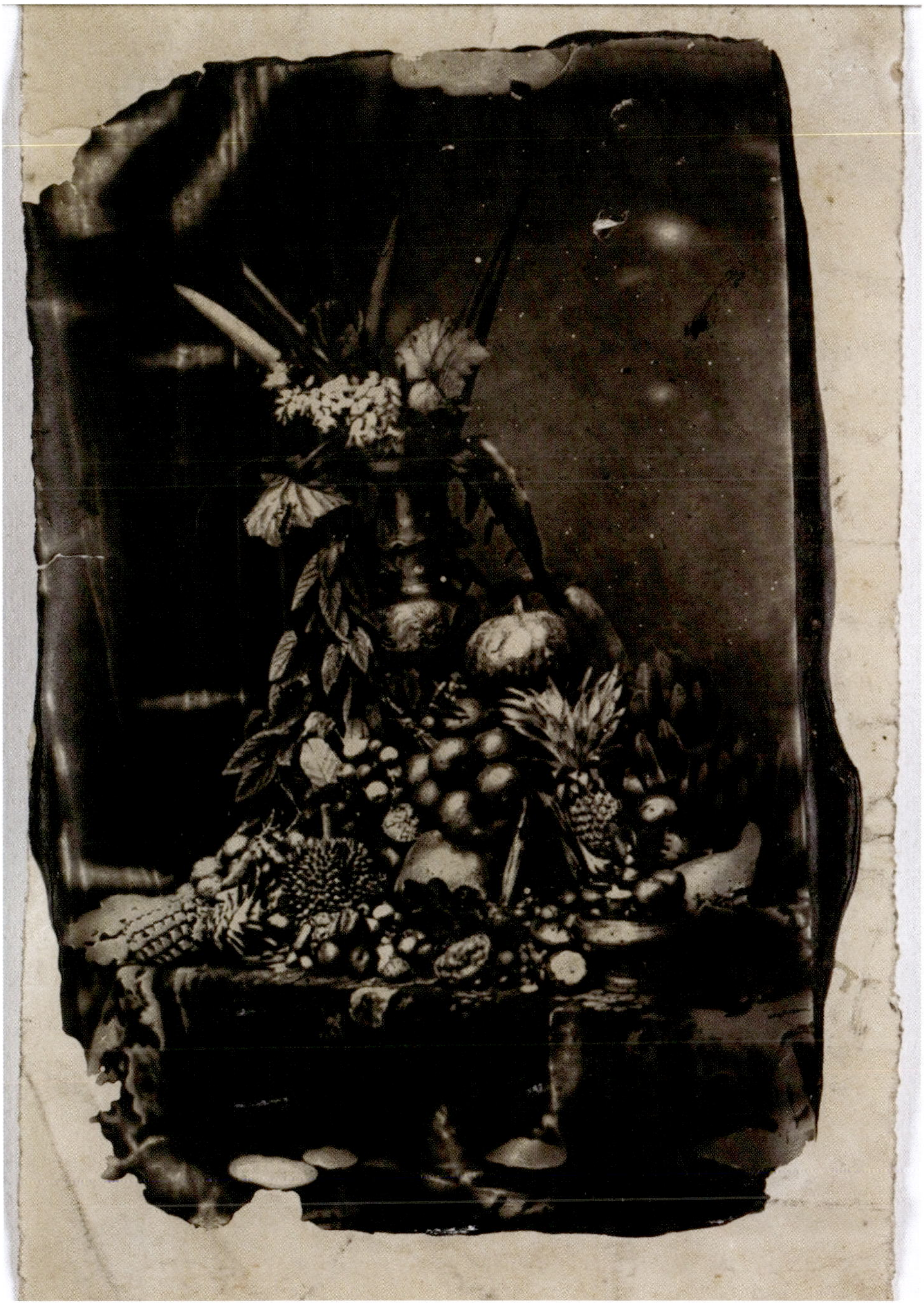

Plate 38. Unpublished very early experimental untrimmed Woodburytype print of still life with fruit. This print displays very little cracking and no peeling compared with plate 37.

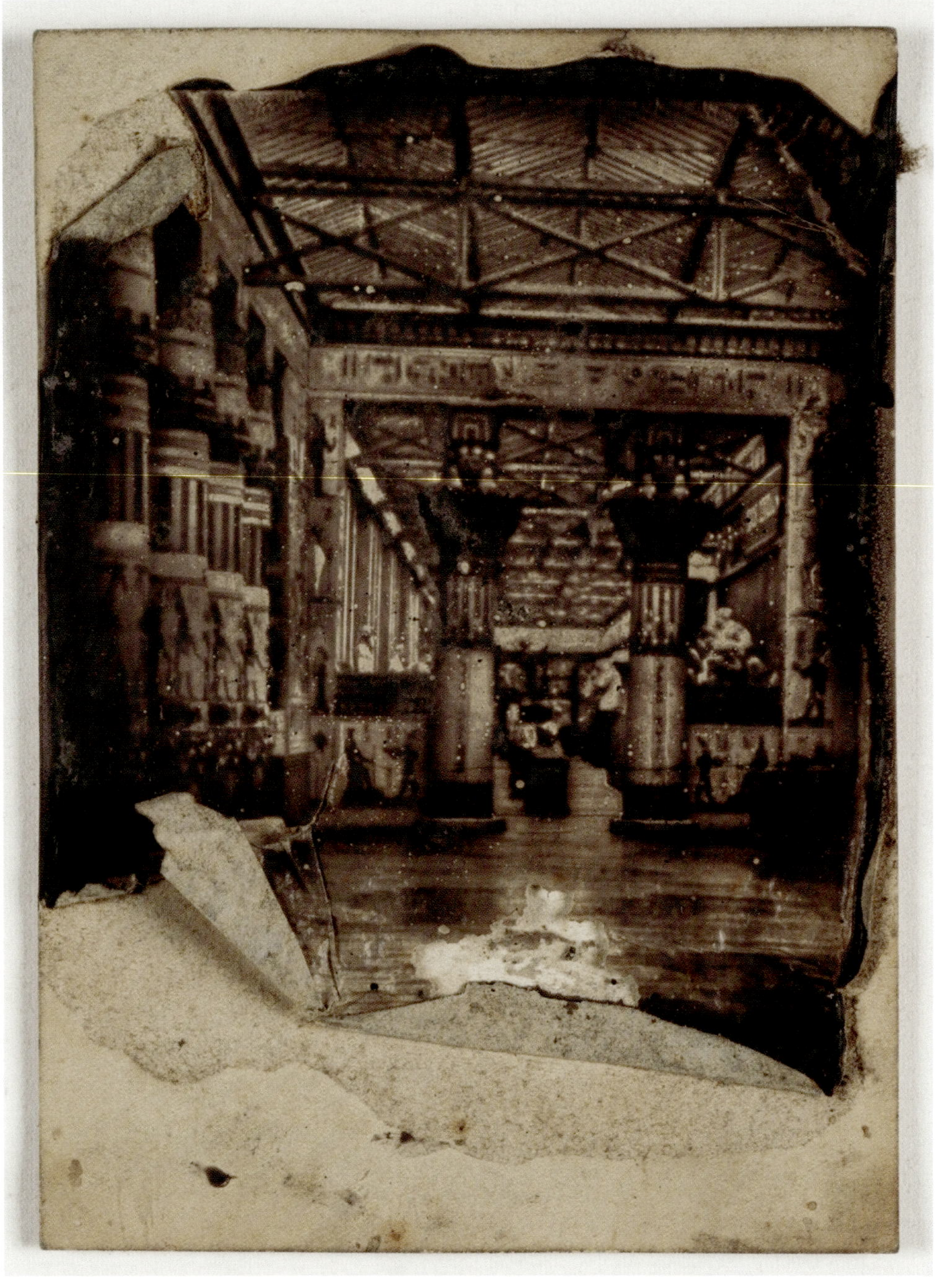

Plate 39. Unpublished very early experimental untrimmed Woodburytype print of the Egyptian Hall at the Crystal Palace, [1866].

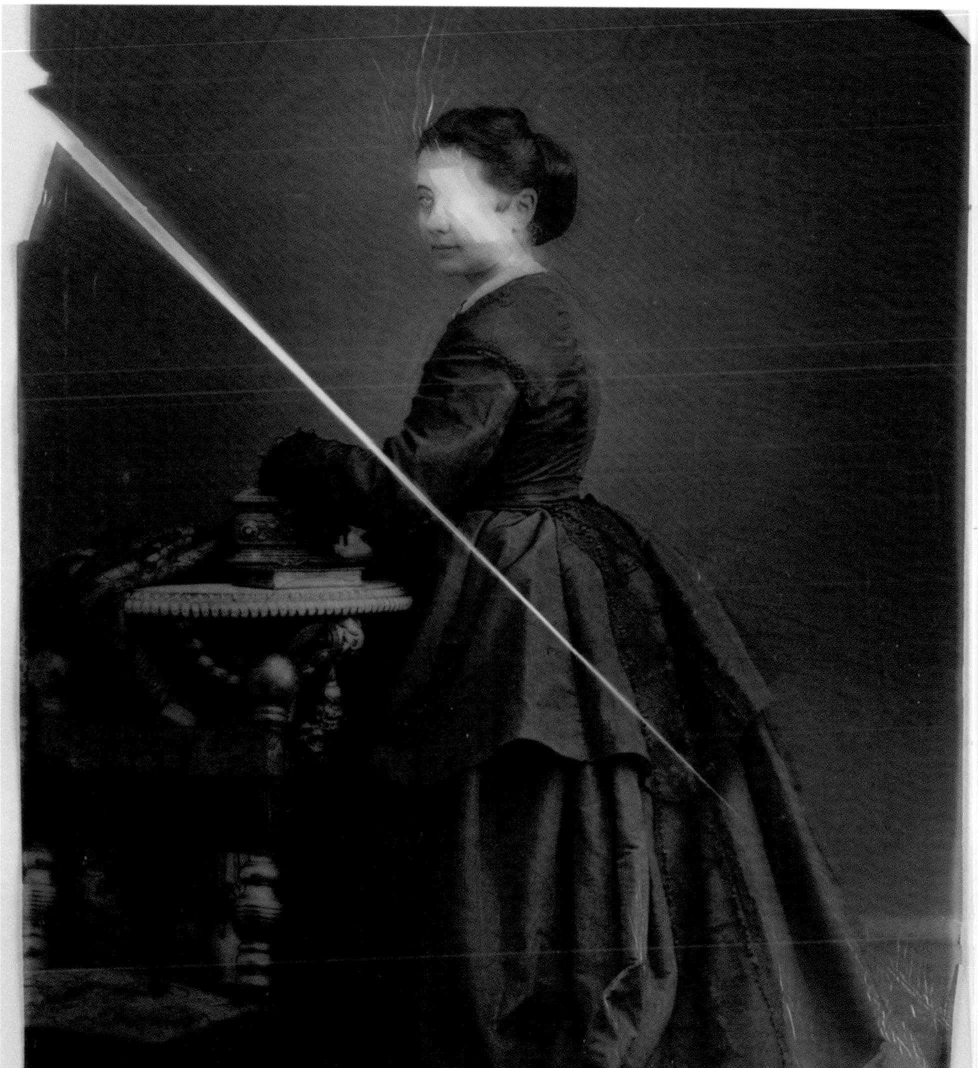

Plate 40. Unmounted Woodburytype relief film of a woman in a dress. While the film displays severe cracking, it is in quite good condition compared with plates 42-44. [The woman in this portrait is possibly Woodbury's wife or one of his daughters.]

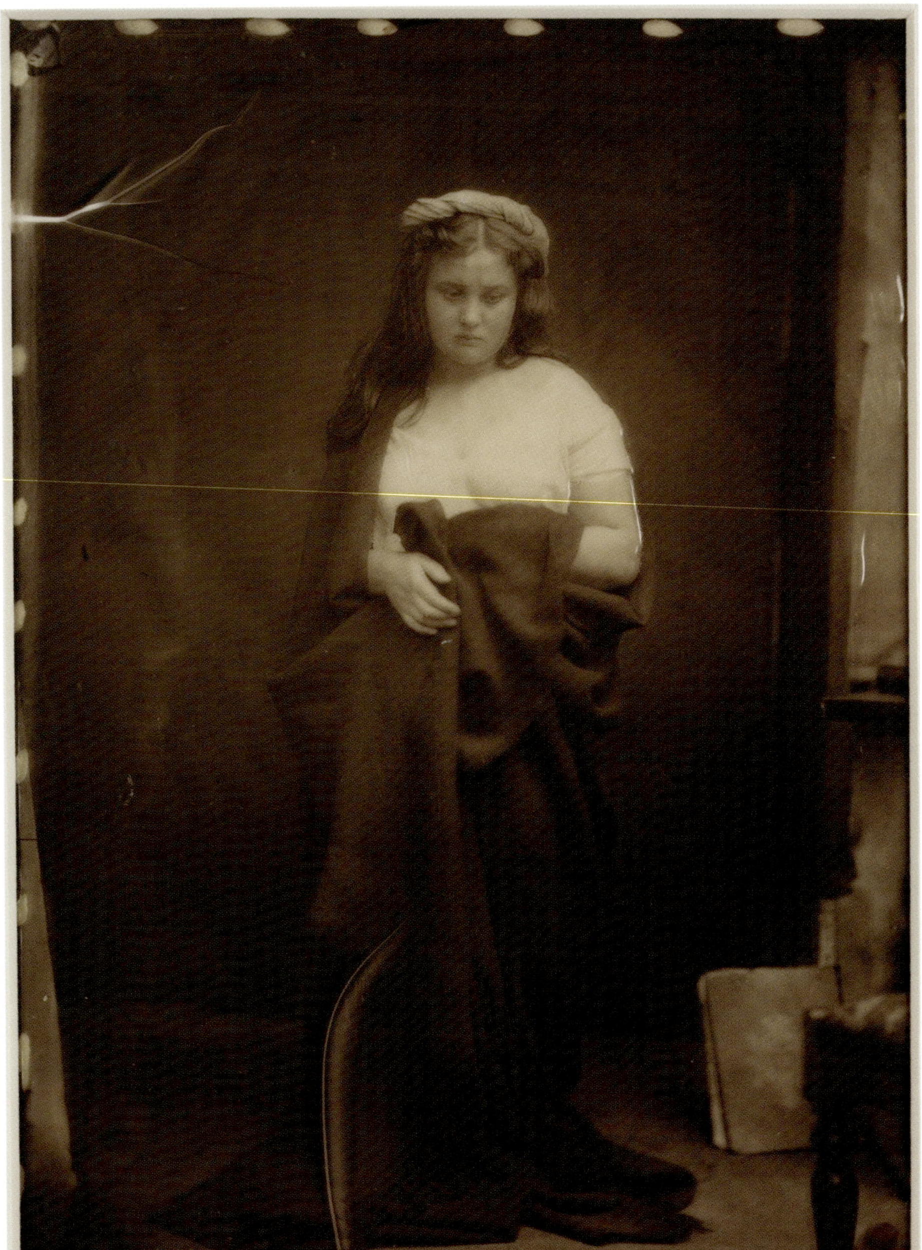

Plate 41. Unmounted Woodburytype relief film of a woman in a dress. This relief film is in similar condition to plate 40. Notice the pattern of dots around the edges which serve as a depth delimiter for the lead plate.

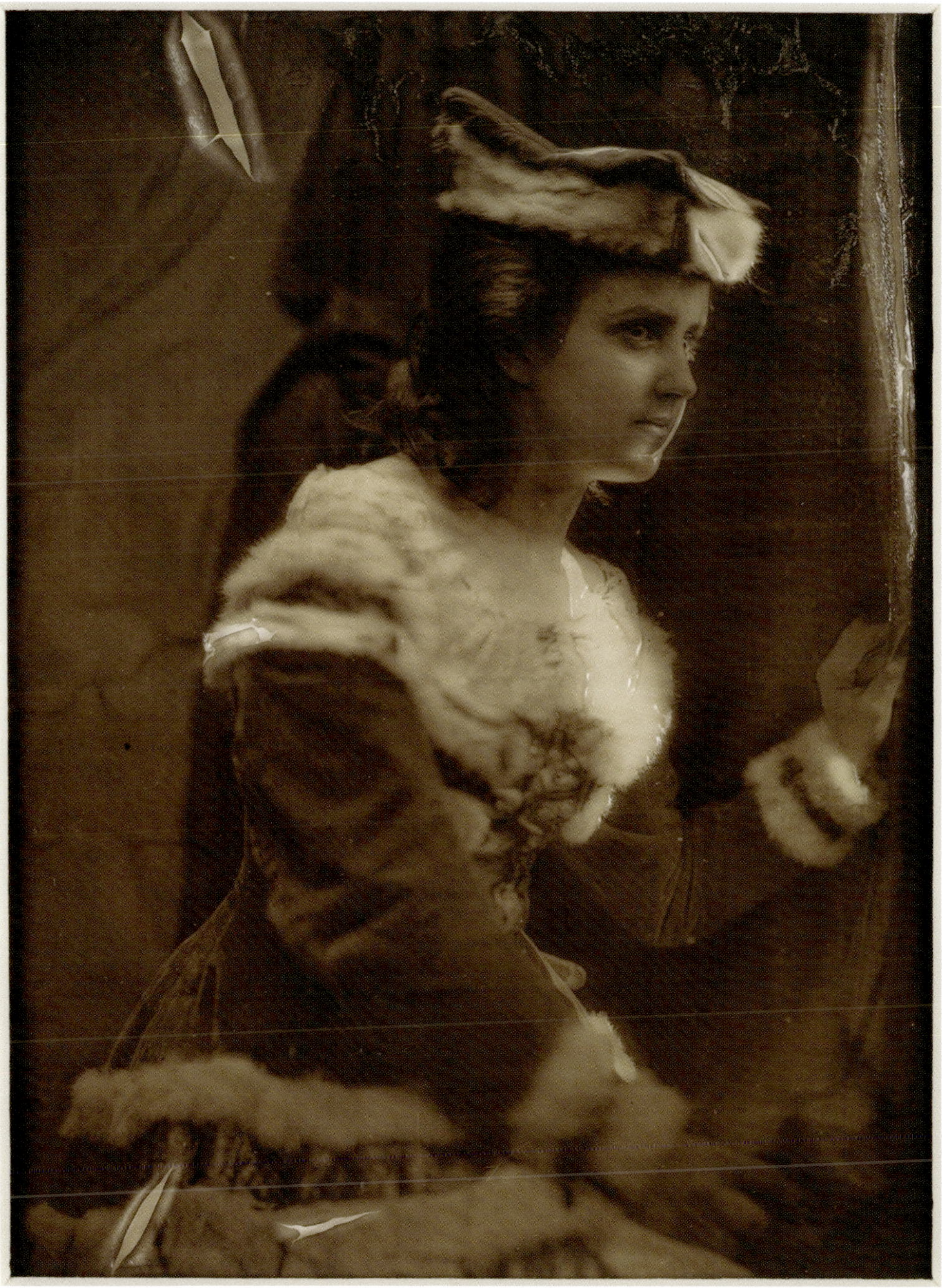

Plate 42. Unmounted Woodburytype relief film of a woman in a dress and fur hat. This relief film is severely cracked, mostly in the thin highlight areas of the gelatin and where high and low values meet.

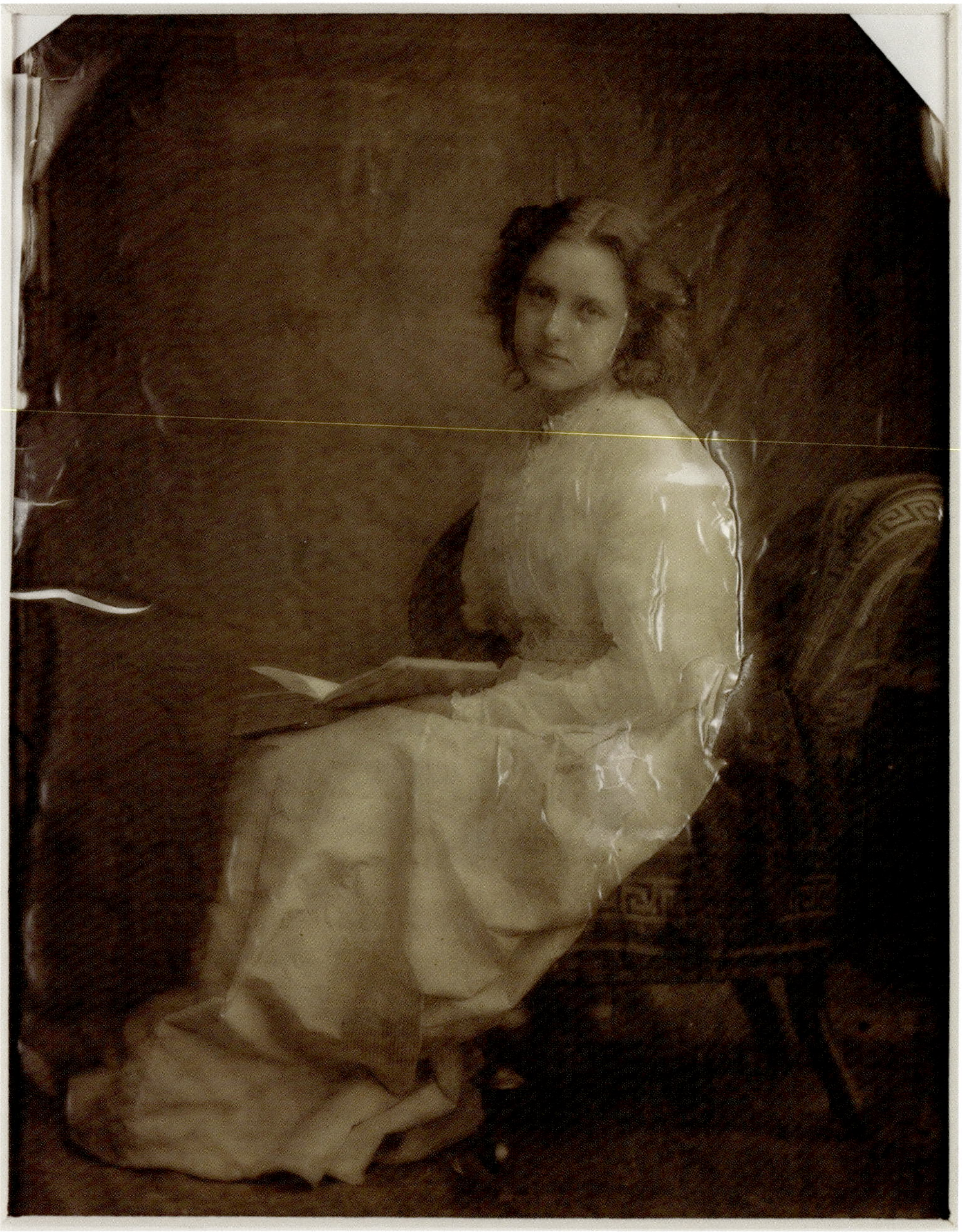

Plate 43. Unmounted Woodburytype relief film of a seated woman in a dress. This relief film is in similar condition to plate 42.

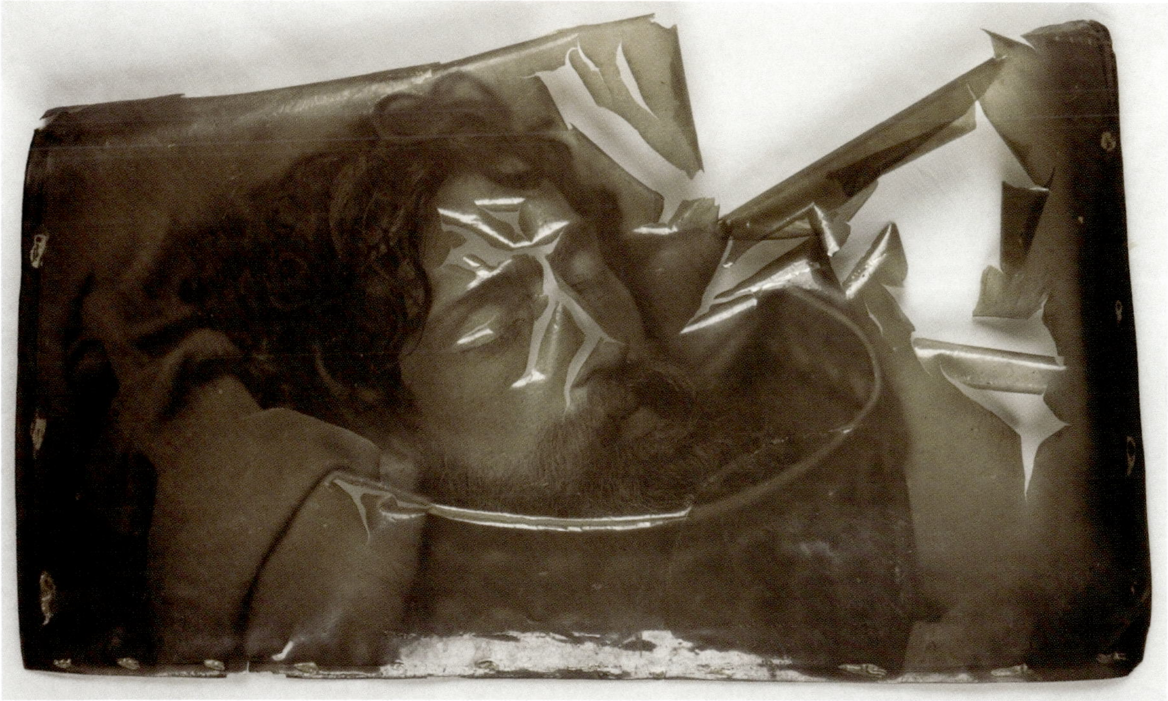

Plate 44. Unmounted Woodburytype relief film of Head of St. John the Baptist, *by O.G. Rejlander. This relief film is severely cracked and curled throughout.*

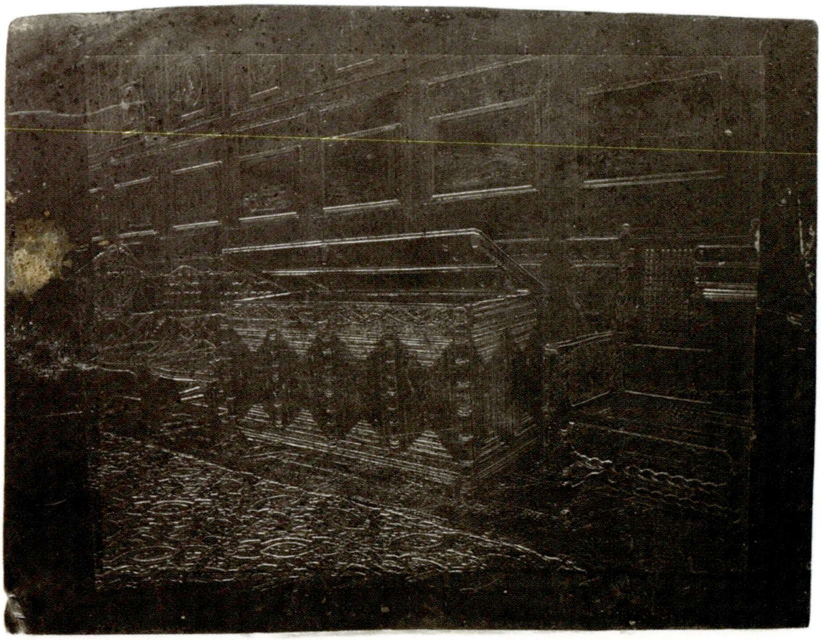

Plate 45. Lead Woodburytype matrix, or printing plate of an interior scene with chest. Although hard to see in this reproduction, the lead plate is in relief corresponding to the density of the negative.

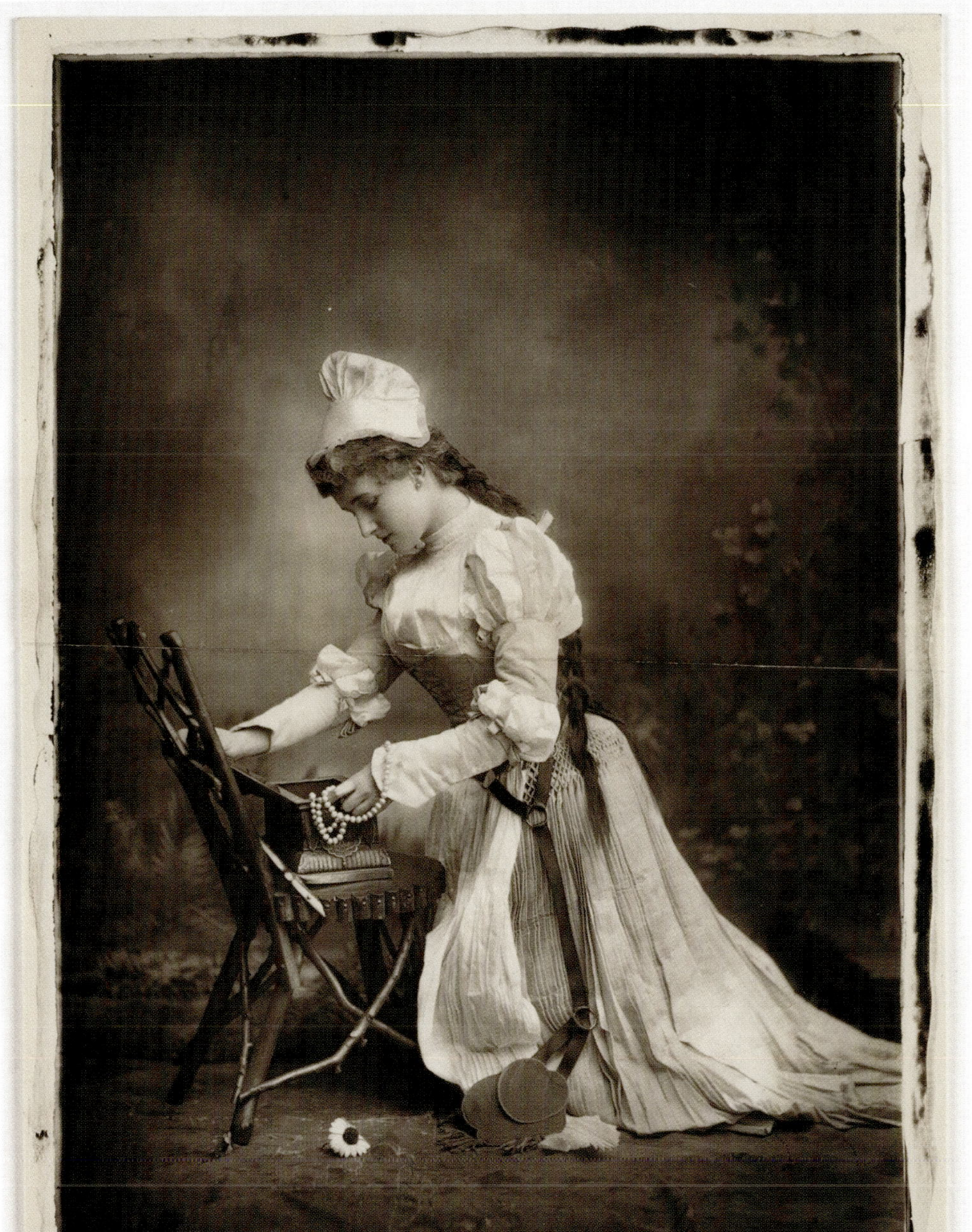

Plate 46. Unmounted Woodburytype print of a woman in a dress. Inscribed on the back "Made at the R.P.S. [Royal Photographic Society] April 1896."

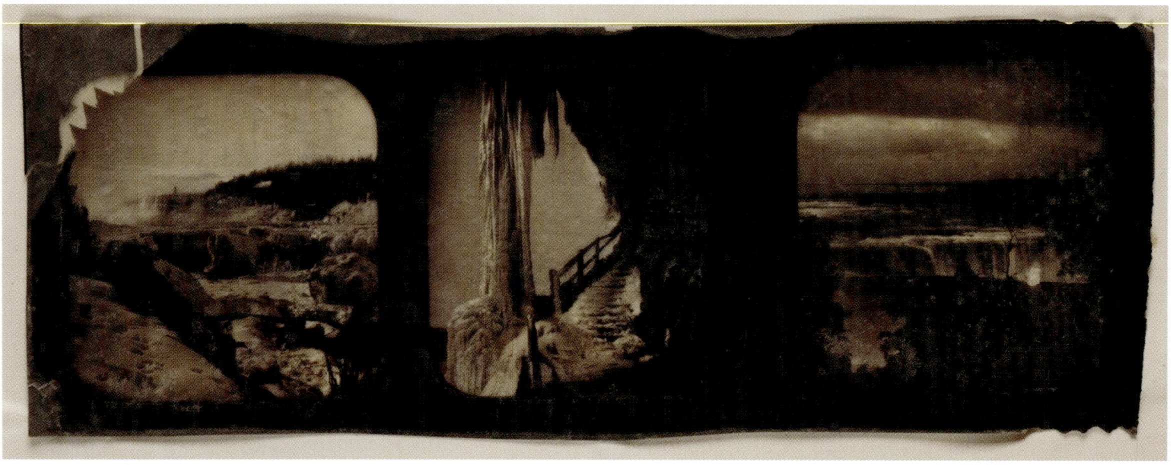

Plate 47. Three Woodburytype lantern slides on a collodion film base [Niagara Falls]. The light discoloration at the top-left corner and around the edges is caused by the collodion separating from the gelatin. Possibly printed by John Carbutt's American Photo-Relief Company.

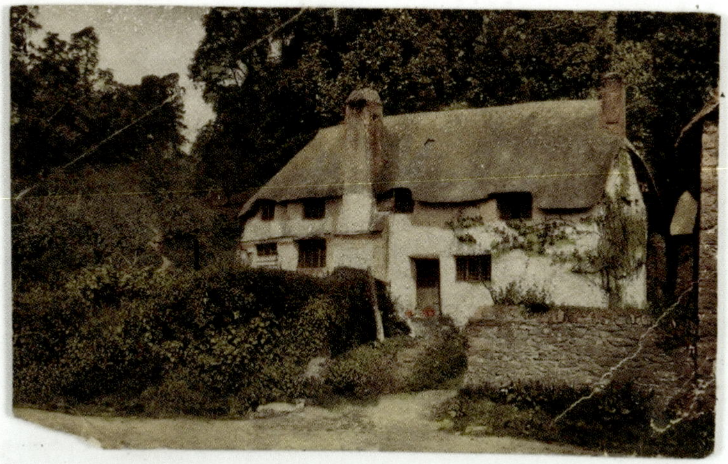

Plate 48a. Hand-colored Woodburytype of a thatched cottage. A normal Woodburytype was printed then transferred to a previously painted paper support.

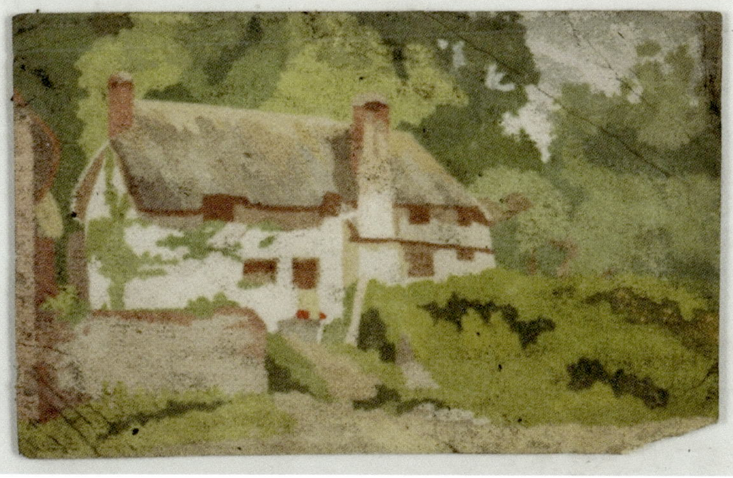

Plate 48b. Same print as plate 48a, seen from the back.

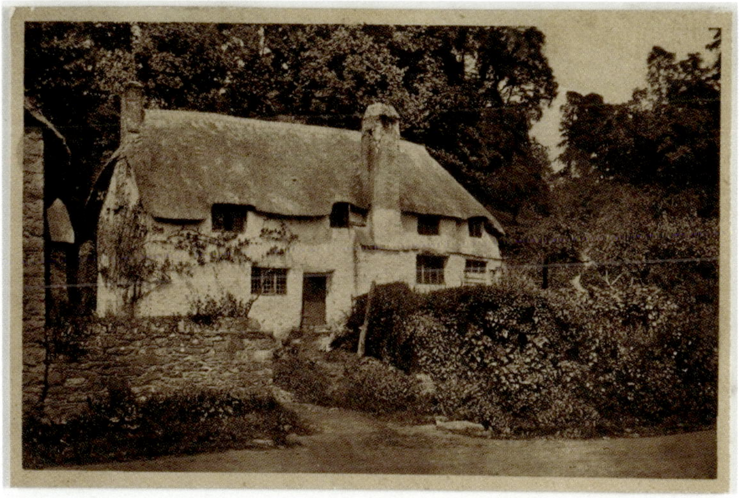

Plate 48c. Normal Woodburytype print. Same image as plate 48a, untransferred.

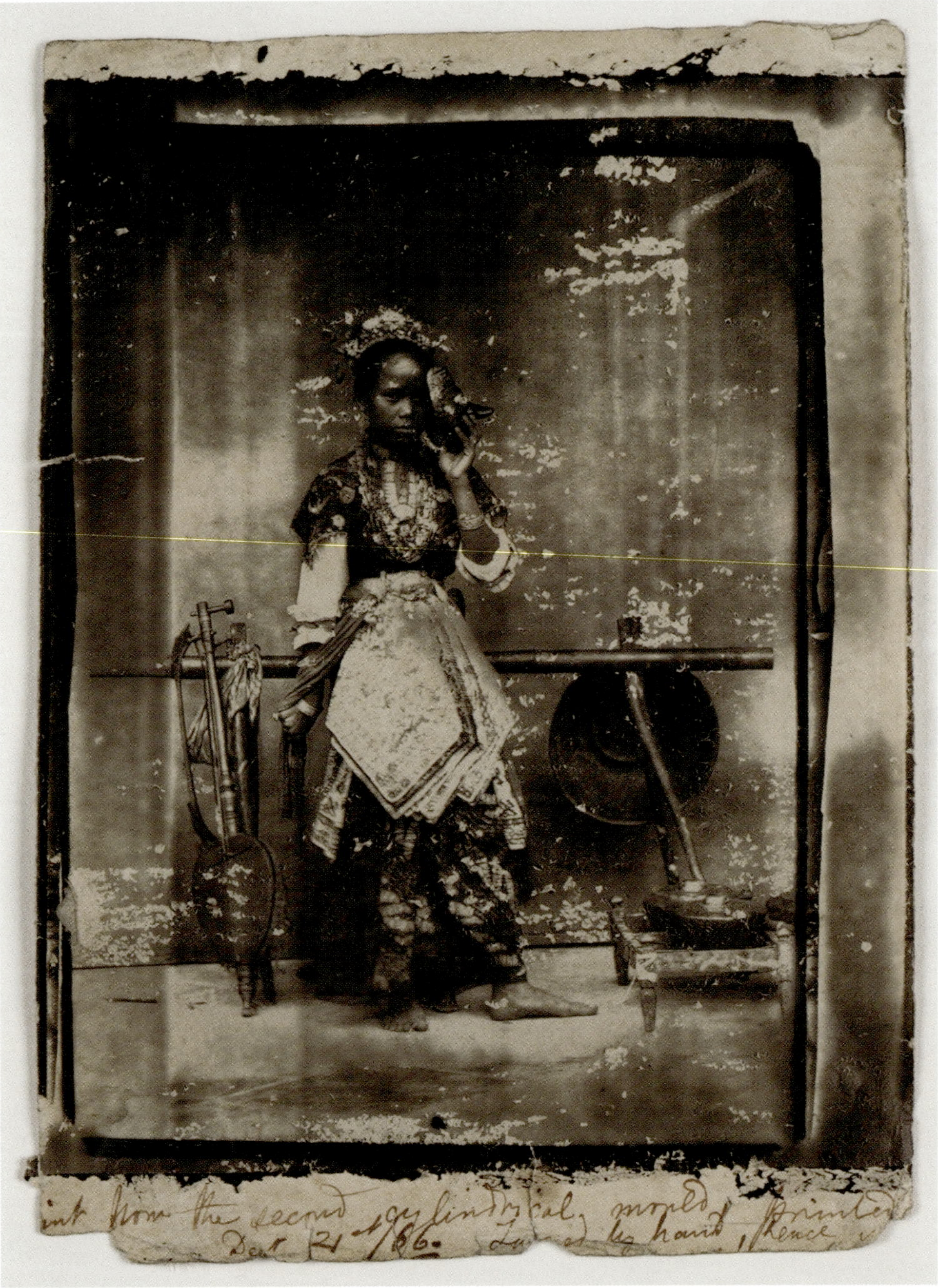

Plate 49. Unpublished untrimmed Woodburytype printed on a roller press of a [Javanese] woman with musical instruments. The inscription by Woodbury reads "Print from the second cylindrical mould printed Dec 21st/66 [illeg.] by hand, hence . . . " and on the back "Early attempt at roller printing from cylindrical mould." Notice the vertical striations caused by the inconsistent pressure of the roller and the horizontal peeling caused by the blade used to scrape the excess gelatin from the mould.

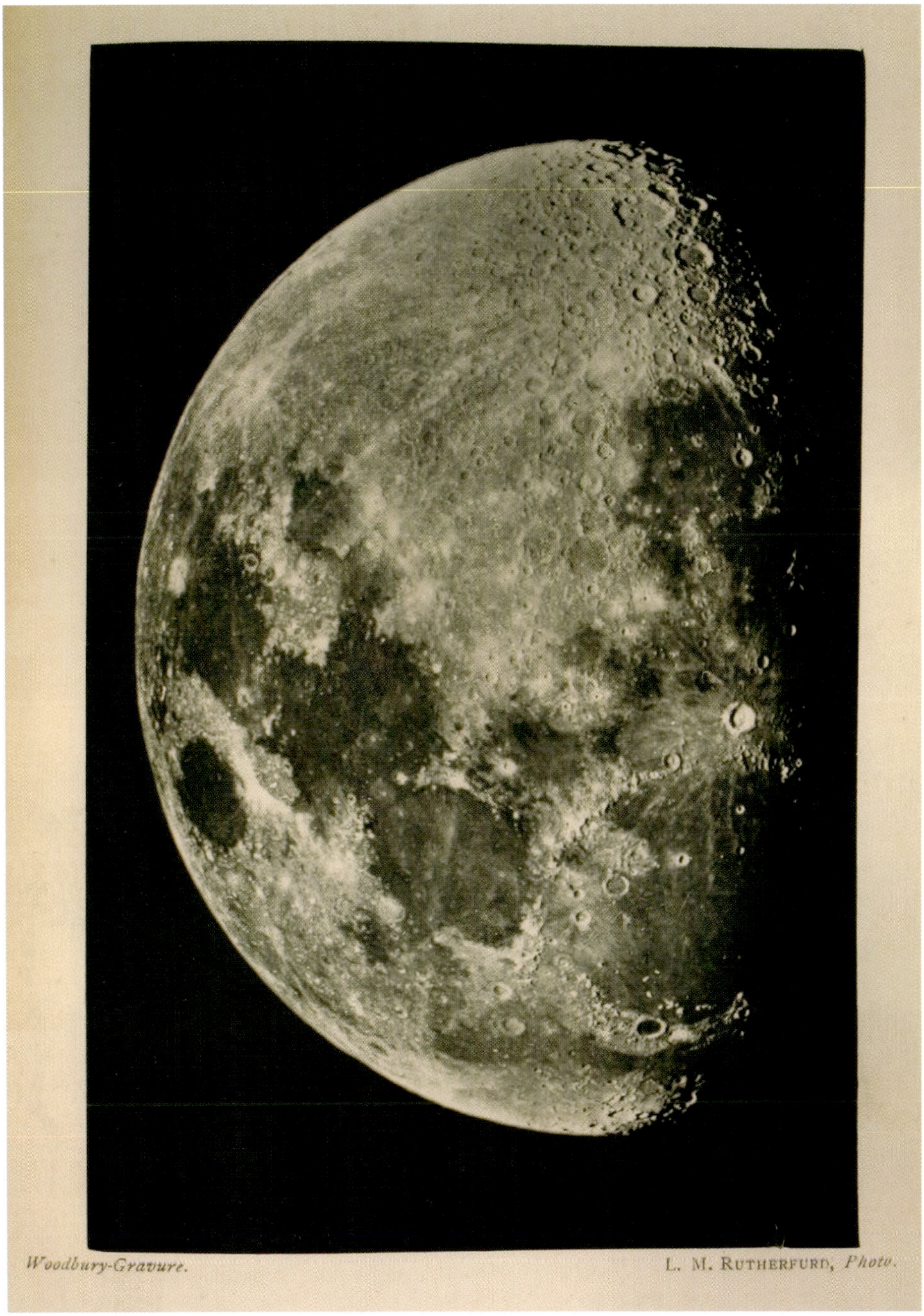

Plate 50. Woodburygravure The Moon-First Quarter, *by L. M. Rutherfurd, printed by the Woodbury Company (Eyre & Spottiswoode), published in* Photography: Its History, Processes, Apparatus and Materials, *1892 by A. Brothers.*

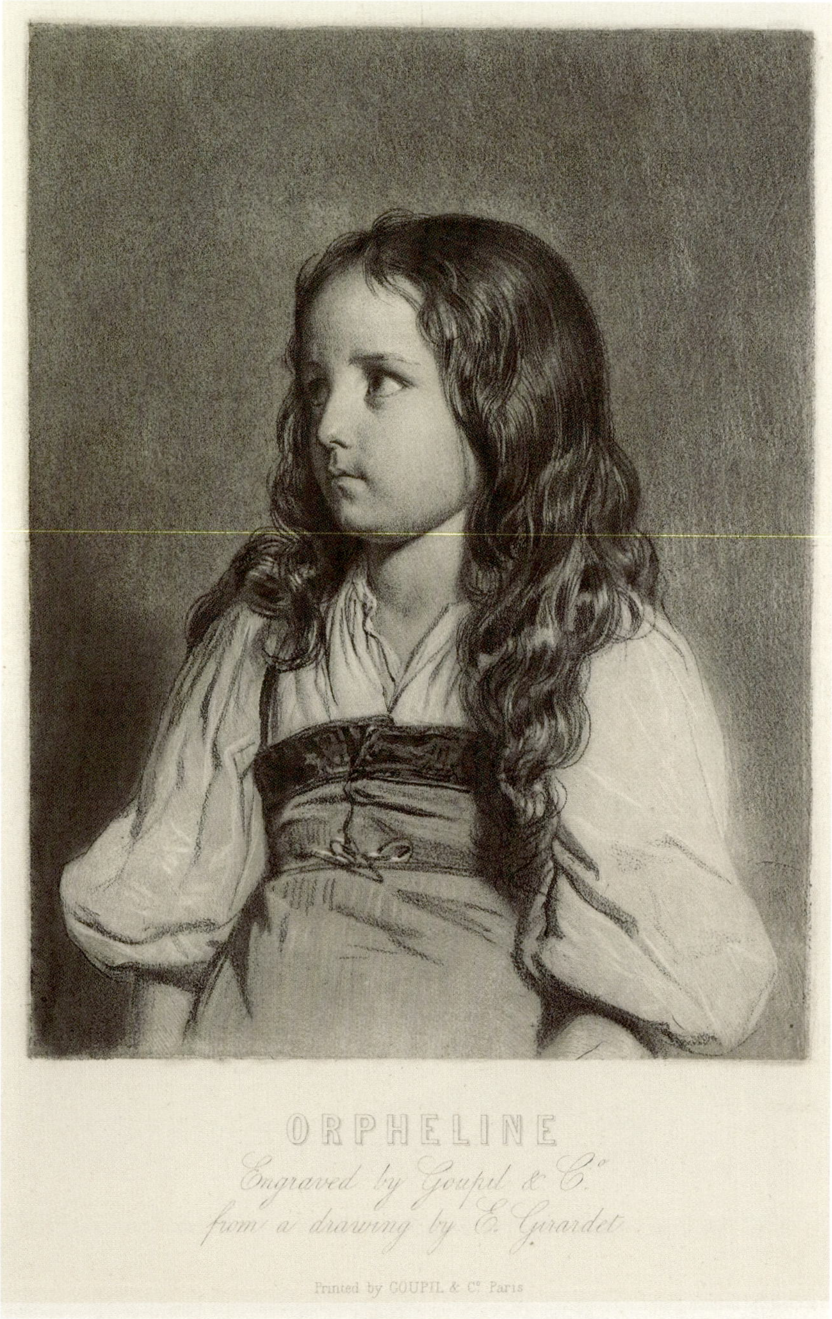

Plate 51. Woodbury's gravure process developed for the Goupil firm, known as Goupil-gravure. Orpheline, *from a drawing by E. Girardet. The plate was made using a Woodburytype relief impregnated with small particles of sand. When this was pressed into the lead plate it produced recesses that would hold the ink and could be printed like a regular etching or photogravure.*

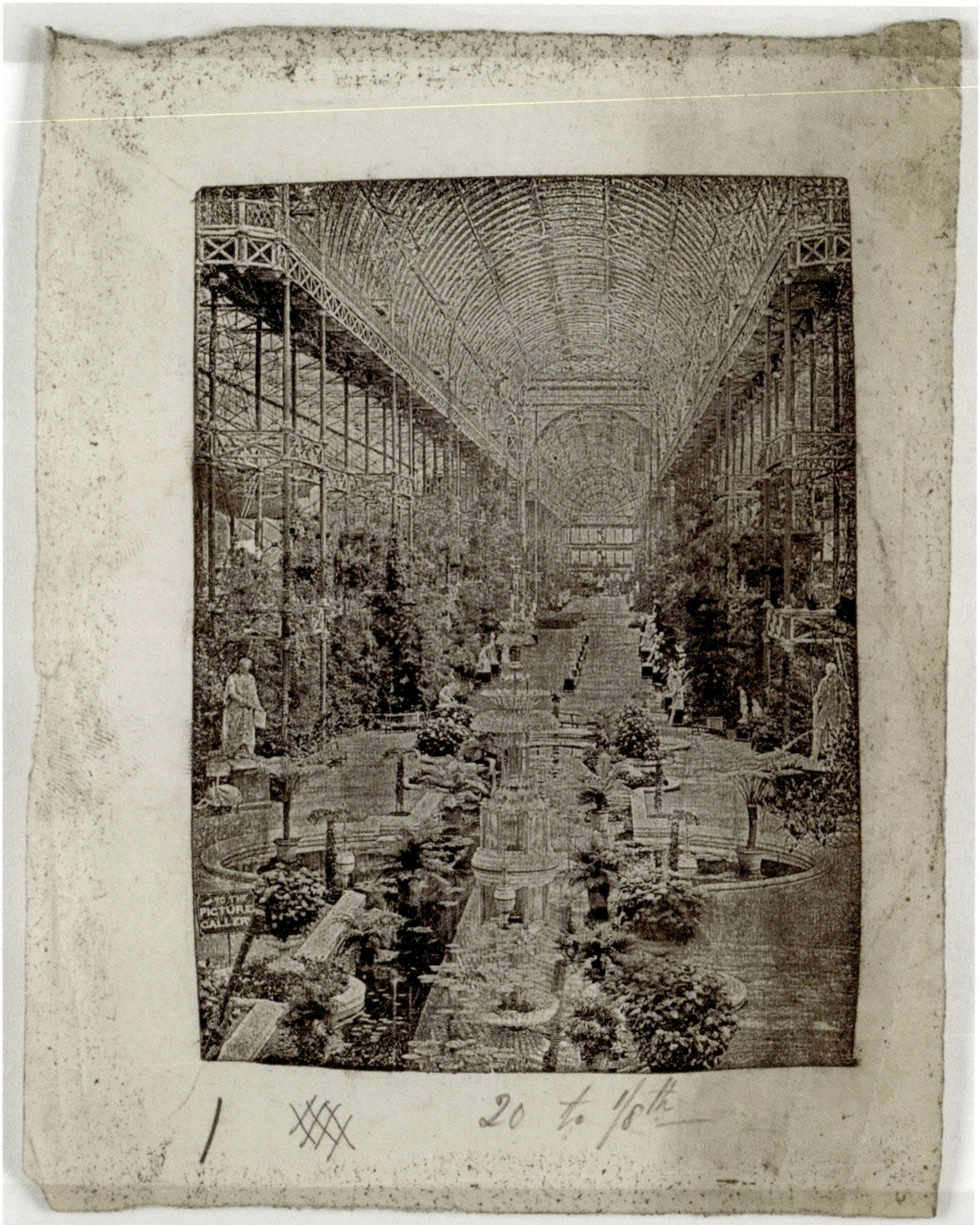

Plate 52. Unpublished experimental print from Woodbury's screen process, printed in ordinary printer's ink. Crystal Palace, [1865]. A negative was printed onto Woodbury-type relief film with a screen interposed, which fragmented the image into tiny dots. When the unexposed gelatin was washed away it left these dots as a raised surface that could carry ink and be printed like a woodblock. See also plates 18 and 20b.

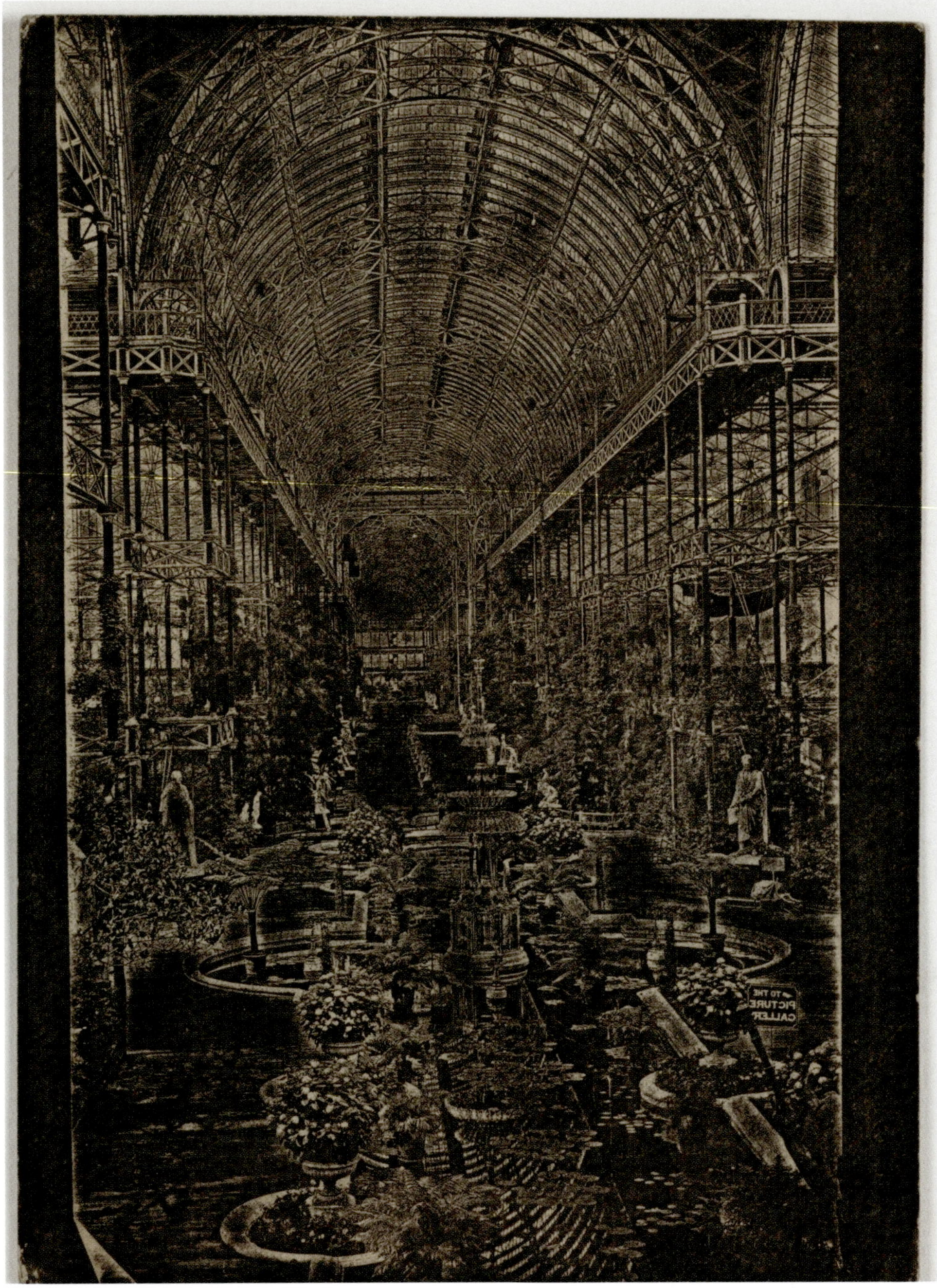

Plate 53. Unpublished experimental print. Same image as plate 52. A regular Woodburytype matrix was surface-inked and printed like a woodblock. Notice the lack of middle tones.

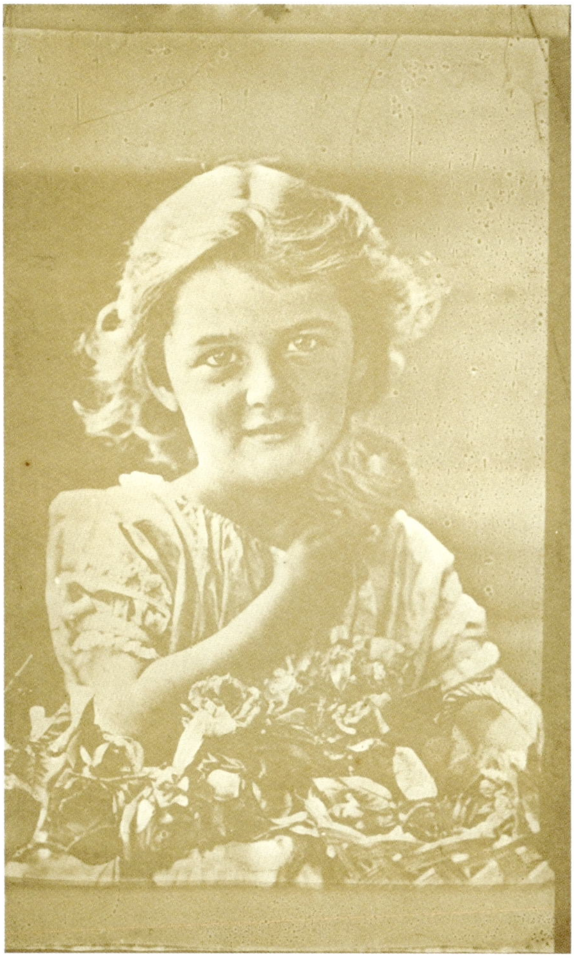

Plate 54a. Unpublished photo-filigrane print, viewed by transmitted light. This print was made using a Woodburytype relief run through a roller-press without ink, compacting the paper fibers (much like a plate mark in etching).

Plate 54b. Unpublished photo-filigrane print. Same print as plate 54a viewed by reflected light. The image is barely visible unless it is viewed by transmitted light.

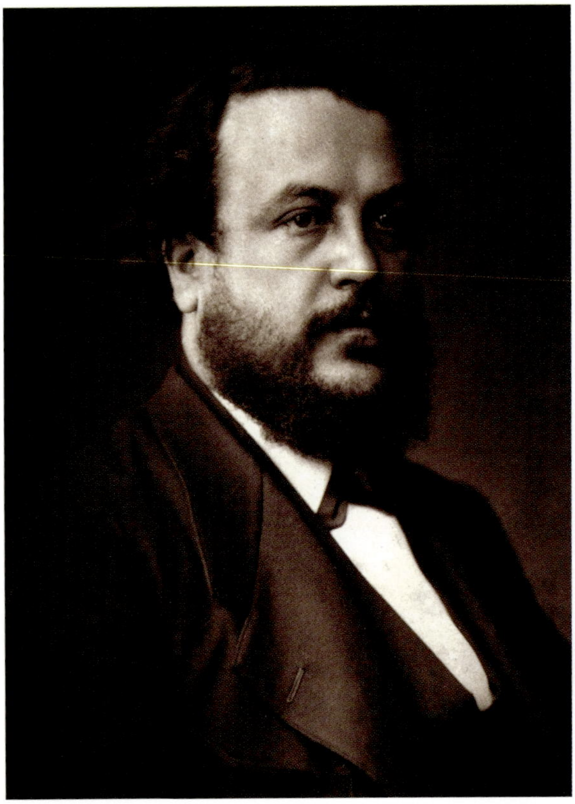

Plate 55a. [Self] portrait of Walter B. Woodbury, made from the Stannotype plate in plate 55b.

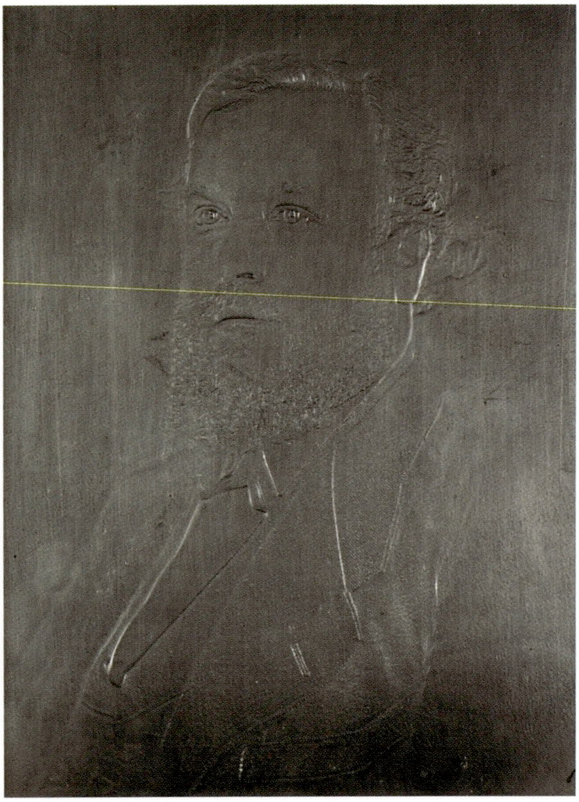

Plate 55b. The Stannotype plate used to make the print in plate 55a.

Plate 56a. Stannotype plate of the same image as plates 55a and 55b, showing the edges and some damage.

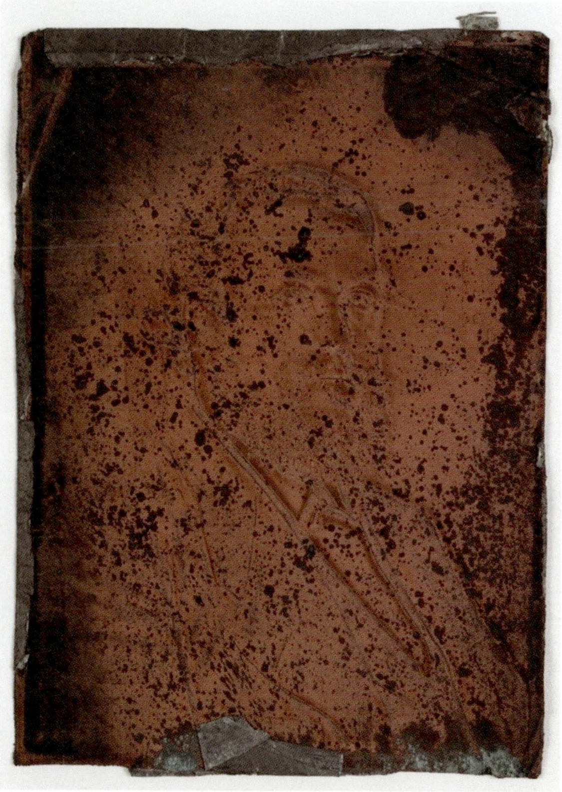

Plate 56b. The same plate as 56a seen from the back. This is an early Stannotype plate made using the electro-type process. The Woodburytype relief film was plated with copper through electrolysis, removed, and then the copper deposit was coated with tin foil. Later, Woodbury did away with the electro-type stage and coated the foil directly onto the Woodburytype relief.

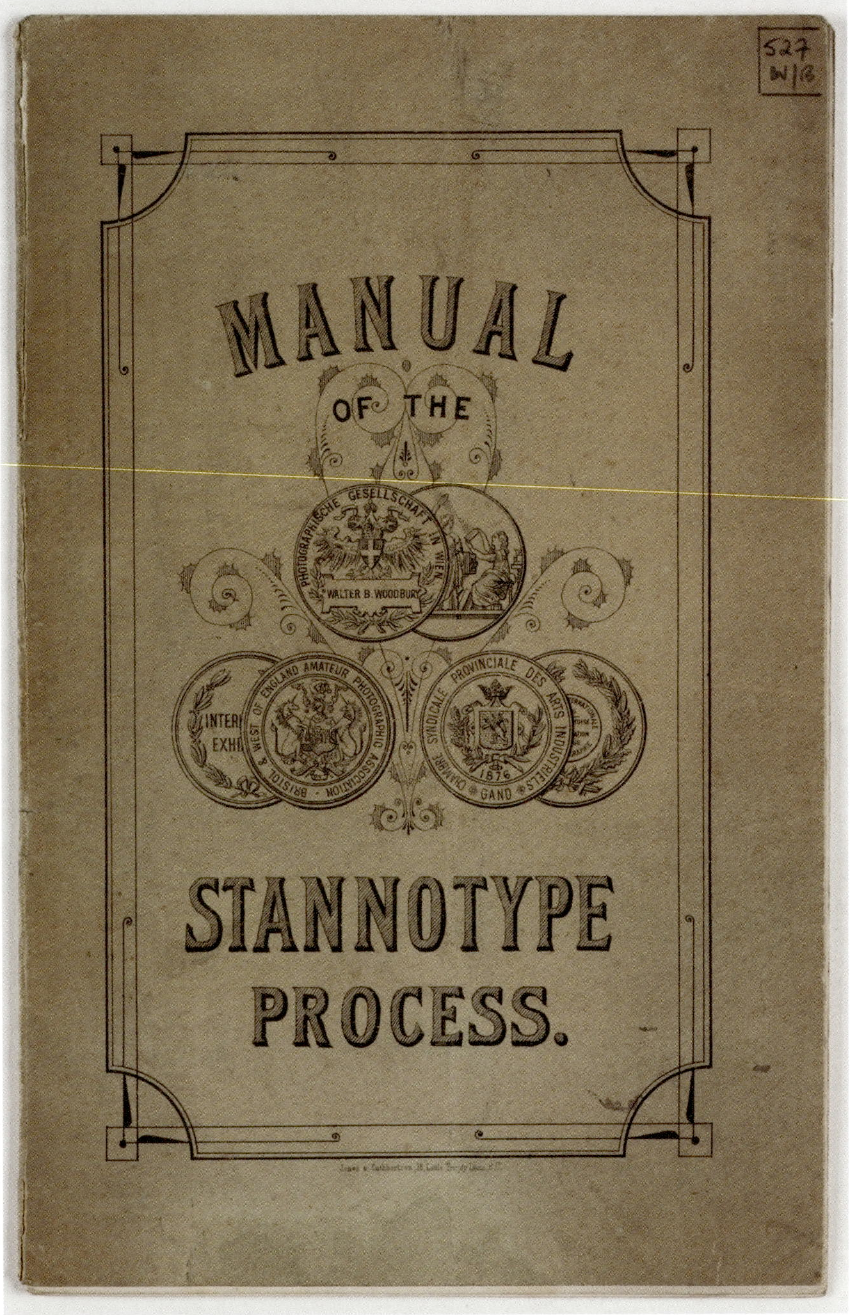

Plate 57. The Stannotype manual, displaying some of the medals the process received.

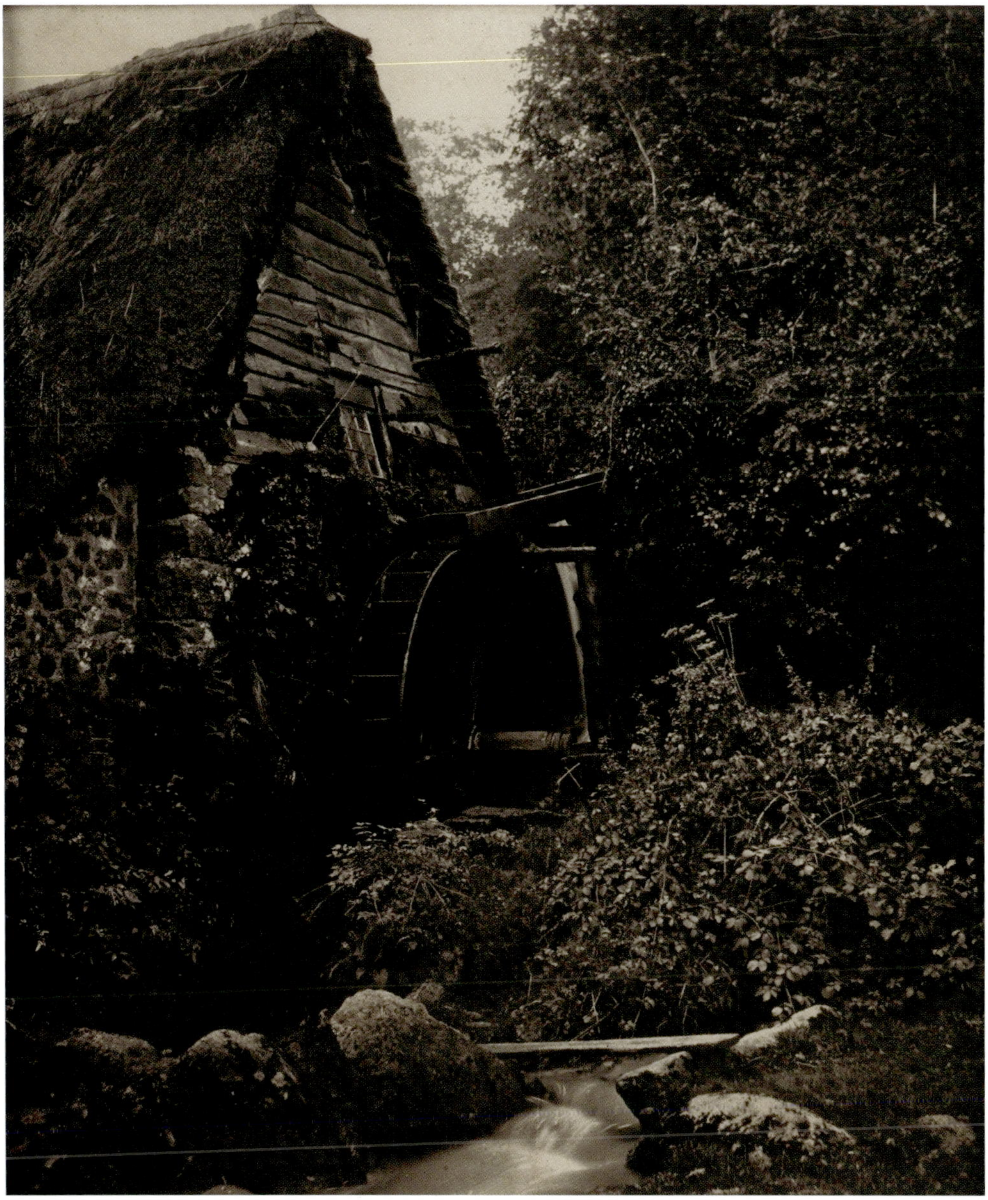

Plate 58. Carbon print, or Mr. Swan's Process, Holy Street Mill *[possibly by J. W. Swan].*

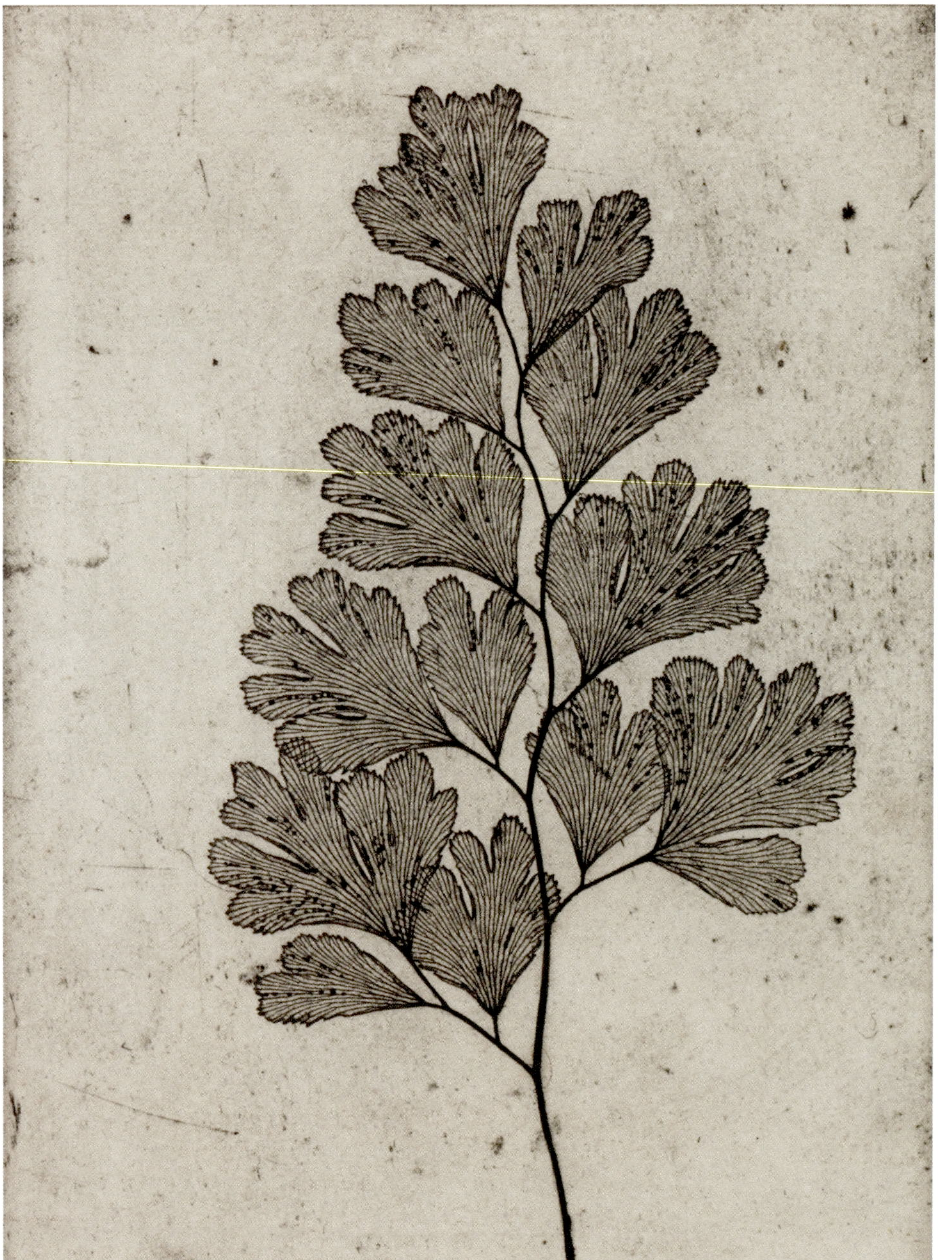

Plate 59. Nature print of a fern leaf. This print was made by pressing the leaf into a lead plate and printing from this like an etching plate.

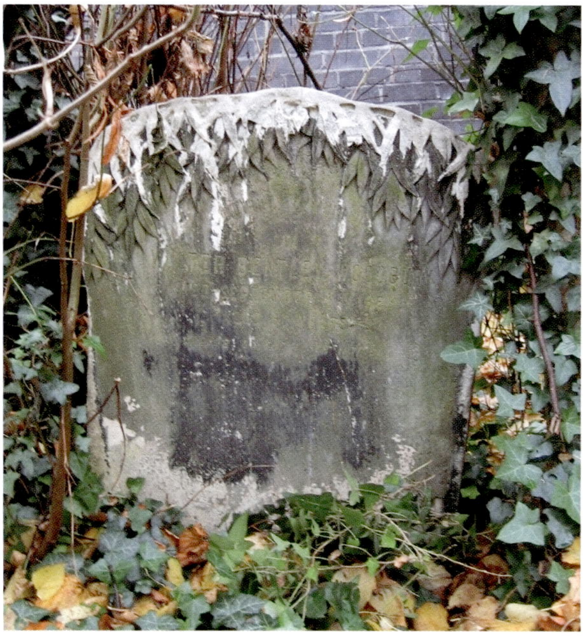

Plate 60a. *Walter B. Woodbury's headstone at Abney Park Cemetery, Stoke Newington, London. It reads "Loving Memory of Walter Bentley Woodbury Died September 4th 1885 Aged 51 Years - Some Men Labour/Others Enter into Their Labour."*

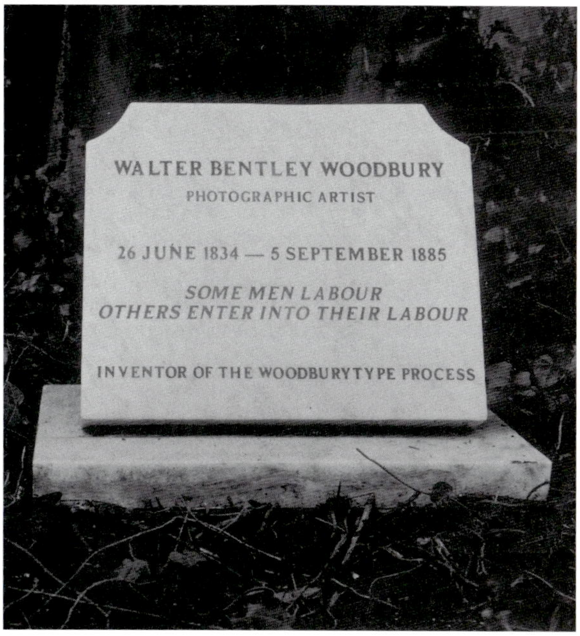

Plate 60b. A new headstone placed by Woodbury's grandniece. It reads "Walter Bentley Woodbury Photographic Artist 26 June 1834 - 5 September 1885 - Some Men Labour/Others Enter into Their Labour - Inventor of the Woodburytype Process."

The underlying principle that made Woodbury's process possible was Mungo Ponton's discovery in 1839 that bichromate salts could be used to produce photographic images.[116] In 1798 Vauquelin discovered that chromate salts become insoluble as they are exposed to light. Building on this principle, Ponton made his first print by soaking paper in potassium bichromate and exposing this under a photographic negative. The salts became insoluble in reverse proportion to the negative and when washed, the soluble salts were removed leaving an image on the paper. These prints were known as chromatypes.[117]

Talbot expanded on these discoveries and found that when the chromate salts are dissolved in gelatin and exposed to light the gelatin also becomes insoluble.[118] This principle also applies when the salts are added to other organic colloids and is the basis of gum-bichromate printing.

Alphonse Poitevin used these discoveries to develop the collotype, which later became the major competitor to the Woodburytype. The collotype uses a sheet of bichromated gelatin, exposed to light and swelled in water, which produces a surface similar to a lithographic stone. When greasy printer's ink is applied to the surface it is rejected by the soluble, water soaked areas, yet is held by the insoluble areas. This is then run through a press to produce an image.

Chromated gelatin was an essential part of many of the early photomechanical processes and contributed to the development of photogravure and photolithography, among others.

— How the Woodburytype Works —

The production of the Woodburytype can be broken down into three major steps. In the first step, a gelatin relief "print" is made from a negative. This is similar to a carbon print and requires the use of light and light sensitive materials. Second, a reverse matrix is made from the relief in a sheet of lead. Third, a print is made from the lead matrix with a solution of pigmented gelatin. A brief description follows.[119]

First and very important, because of the sensitivity of the materials to be used, a strong, contrasty negative, preferably of collodion, is selected. Then a sheet of bichromated gelatin is prepared by mixing together a solution of white gelatin and potassium bichromate, coated onto a sheet of glass and placed in a desiccating box until dry *(fig. 3)*.

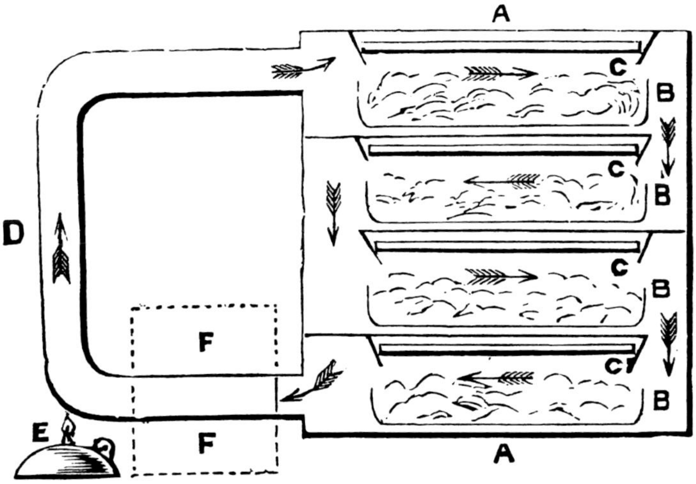

Figure 3. Desiccating box used in the Woodburytype process.

The negative is then contact printed onto the sheet of bichromated gelatin. It is exposed to the sun for anywhere from five minutes to an hour, which produces a rather weak image by turning the bichromates dark. An actinometer is required for accurate exposure.[120] The bichromate acts on the gelatin by turning it insoluble in direct proportion to the amount of light it receives. This is a positive version of the image.

This film is then developed in hot water, which washes away all the unexposed portions of the gelatin, producing a relief in reciprocal proportion to the density of the negative. In parts of extreme highlight (where the negative is most dense) the relief will be thinnest and in the parts of darkest shadow (where the negative has little developed silver) the relief is thickest. It is then dried and hardened; the resulting sheet is only 1/2000th inch thick.

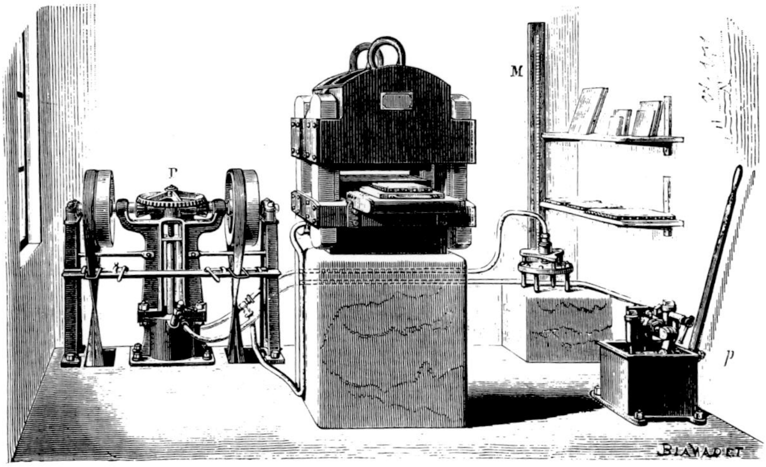

Figure 4. Hydraulic press used to make Woodburytype lead plates.

Second, after drying it is pressed into a sheet of lead under immense hydraulic pressure. The press must be able to deliver 5,000 pounds of pressure per square inch; for an 8 x 10 inch plate a pressure of 400,000 pounds is needed *(figs. 4 and 5)*. This produces the printing matrix. It is a negative image and the lead plate has indentations corresponding in depth to the density of the original negative. It looks somewhat like a reverse bas-relief *(plate 45)*. The hydraulic press and lead plate was necessary because an absolutely true, flat surface was required to obtain the correct gradation in the print. It can then be used as a printing plate.

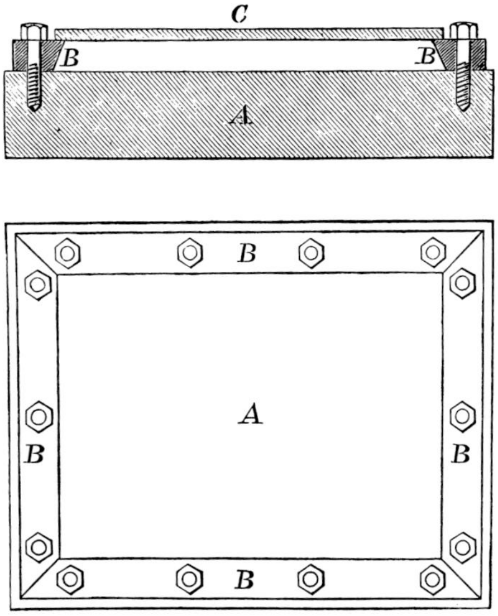

Figure 5. Platen used to hold lead plates in place while in the hydraulic press.

Third, paper is specially prepared by coating it with a series of shellac and gelatin sizes, then running it through a calendering press to make it absolutely smooth. This is necessary as any imperfections in the paper surface will affect the density of the ink in the print.

Then, a unique ink is manufactured using a mixture of liquefied gelatin and pigment. It is poured into the depressions in the plate, combined with the paper support and put under slight pressure in a specially devised press *(figs. 6 and 7)*, similar in appearance to a book press. After about five minutes the gelatin sets enough to be removed from the press. It is then hardened in a chemical bath, dried and trimmed to size. These prints were tipped-in to publications or pasted directly to individual cabinet cards.

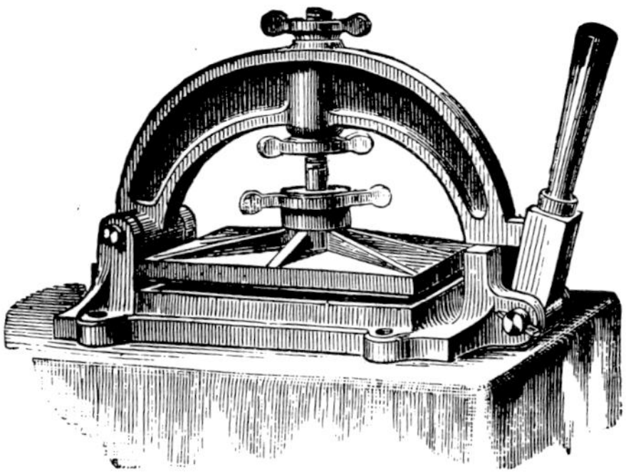

Figure 6. An early model Woodburytype printing press.

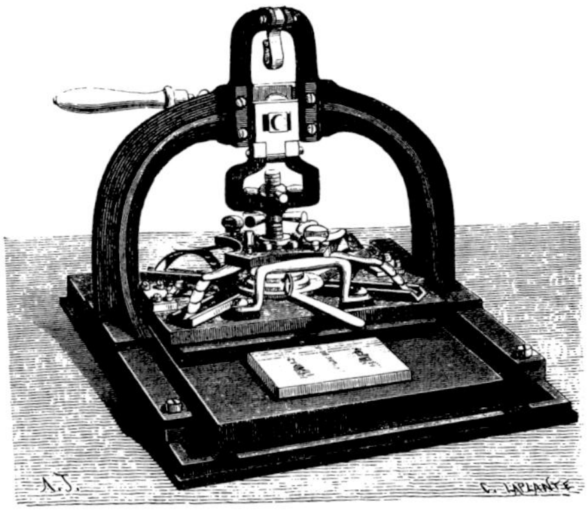

Figure 7. A later model Woodburytype printing press, showing the printing matrix, or plate.

Since a single gelatin relief could yield about six matrices, most shops were equipped with rotating tables *(figs. 8 and 9)* holding six presses. One operator could run this by himself. He started with one press, turned the table to the next and by the time the first press returned the print was ready to come out.

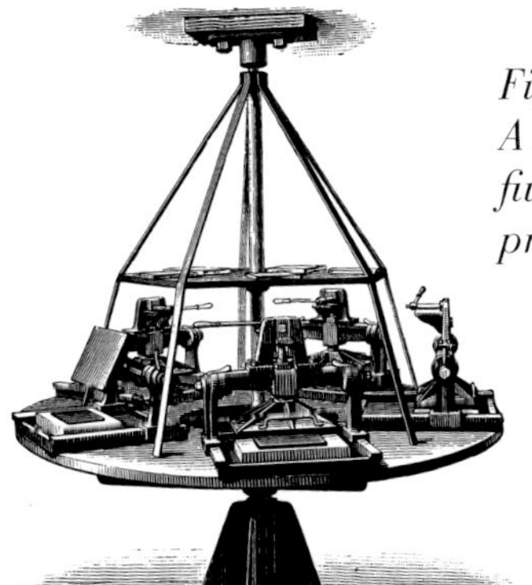

Figure 8. A turntable holding five Woodburytype printing presses.

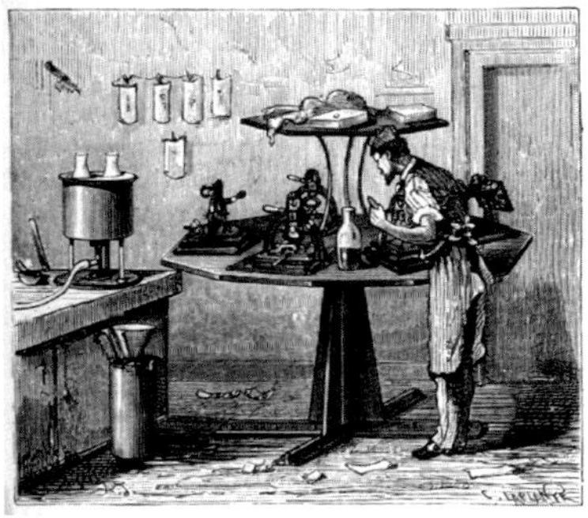

Figure 9. Printer printing at a turntable holding six Woodburytype printing presses. On the table and the counter on the left are bottles of printing ink.

The process looks like this: *(fig. 10)*

original scene ->
camera negative ->
relief positive ->
matrix negative ->
print positive

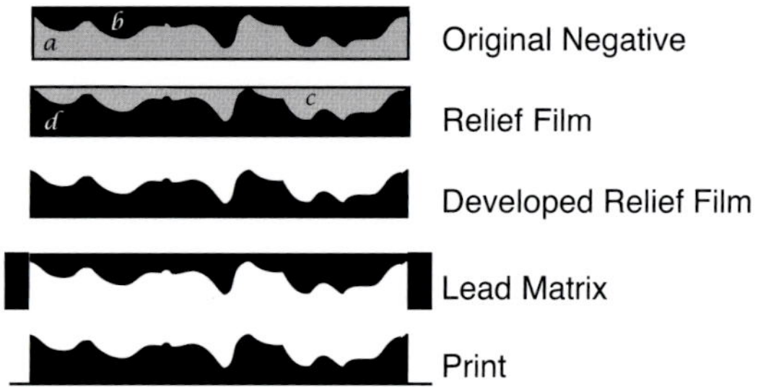

a = unexposed negative
b = developed silver
c = unexposed relief film
d = insoluble gelatine

Figure 10. The steps of the Woodburytype process.

The process is very flexible in its applications. Because it is a contact printing process, if an image size is to be changed from the original camera negative, an interpositive is first made, from which a correct size negative is enlarged or reduced.

The peculiar properties of the semi-transparent ink made the Woodburytype ideal for the production of lantern slides.[121] If the paper base is substituted with glass and the ink mixture made a little more dense to compensate for the transparency, a slide of great detail and permanence could be made.[122] This outgrowth of the process accounted for a large part of its commercial production *(plate 47)*.[123] It also allowed for hand-coloring *(plates 48a-c)*.

◆

What happened to this most beautiful, efficient, hi-fidelity process? Despite its almost complete acceptance by consumers and printing firms, the Woodburytype suffered from a few major problems that led to its eventual disuse. With advancing technology, printers frequently adopted faster and cheaper means of reproduction over superior means that required more expense. The Woodburytype fell victim to this pattern.

One problem was the initial outlay for the special equipment. Not only was the hydraulic press required to produce the matrix, but other pieces of special printing equipment were needed to produce the finished print. As the size of the hydraulic press determined the size of the final print, anything larger than about 8 x 10 became prohibitively expensive and exceeded the cost of silver prints. Carbon prints and collotypes, on the other hand, could be made at virtually any size, and since the demand for large prints was minimal, these processes were frequently used instead. Woodbury had made larger prints, however, and as the technology for hydraulic presses advanced these became more of a possibility.

Another major problem was technical, and it could not be overcome with more money or advances in outside technology. The Woodburytype required a special ink to produce the image. Ordinary printer's ink, which is oily, did not have the semi-transparent properties needed to produce the subtle tonal range of the Woodburytype. So, these plates could not be incorporated with text to make a homogenous printed page. The image had to be printed separately from the text and tipped-in. The ink also was responsible for one flaw that was never overcome in the process, which was its lack of ability to produce large, even areas of white.

The profit from increased production would have made these problems irrelevant, if it were not for continual advancements in text printing. Much like the Daguerreotype, the Woodburytype's quality couldn't compete with the accelerating demands for speed and quantity. Although some firms were reportedly turning out over a thousand Woodburytype prints per day, it was still laborious. The output just couldn't keep up with that of a newspaper. Newer, faster technologies gradually displaced the Woodburytype and by 1900 only a handful of publishers continued to use it.

Decades earlier when the *Times*, London, installed its first rotary press, text printing reached unheard of quantities. In only a few short hours, tens of thousands of copies could be printed.[124] At first this was limited to newspaper production, but other areas of the publishing industry soon started adopting faster methods for text printing. Because Woodburytypes had to be tipped-in, this limited their inclusion to special edition books and journals. By the time the color rotary press came into use, the Woodburytype was completely incompatible. Scientific experimentation continued until a photographic image could be included in all printed matter.

Woodbury was acutely aware of the drawbacks in his process. He knew that printing press technology was advancing and would soon surpass the capabilities of the Woodburytype if it didn't change. Over the years he made several attempts to modify or adapt his peculiar relief printing method to keep up with the demand.

The rotary press occupied most of Woodbury's attention *(plate 49)*.[125] He took out no less than five patents that reflected his researches in this area. This was the future of printing—an image that could come from the fast moving roller, surface inked and included on the same page as the type. Just as the photogravure adapted as the roto-gravure, the Woodburytype would have to adapt if it were to continue to be practiced. None of his attempts at a rotary process succeeded, however, because the Woodburytype requires a period of several minutes under steady, even pressure for the image to form. A fast moving rotary press will deliver pressure for a very short amount of time in one place before the paper advances, never giving the gelatin a chance to set.[126]

The simplest variation, the Woodburygravure *(plate 50)*, is actually a misnomer and has nothing to do with the other gravure processes.[127] Why exactly it was given this name is not known—possibly because it has a duller surface than a true Woodburytype, more like a photogravure.[128] It actually consists of a normal Woodburytype print that is cropped and transferred from its original support, while wet, to a second piece of paper. It could be transferred directly to a book page, avoiding some of the labor of tipping-in.

Another modification, which was more successful, was the inclusion of grain. An intaglio process like the photogravure depended on recesses in a copper plate to hold the opaque ink that traditional printers used. Woodbury had taken out several patents with this in mind and in 1870, at the urging of the print firm Goupil, developed a way to add grain in the first relief or the lead plate by using small particles of sand, producing the needed recesses.[129] Woodbury traveled to the Goupil plant in France and worked directly with its manager, Roussellon, to work out the details of the process.[130]

The *Photographic Art Journal* published the first example of Woodbury's gravure process, a reproduction of *Orpheline* by the French artist Girardet *(plate 51)*.[131] The journal was so enthusiastic about the process that it printed an article devoted to its merits and workings.

Goupil later became famous for their gravure prints and the process was known as the Goupil-gravure to distinguish it from the photogravure process later developed by Karl Klic. Most of Goupil's prints, however, were not inscribed with the process name and it is impossible to determine whether they continued to use Woodbury's process or switched to the Klic method.

A similar modification produced common relief printing blocks, which were adaptable to the text press.[132] These could be made from original line drawings or a regular photograph with a net-work pattern superimposed on it and surface inked, like the halftone process.[133] Quite a number of Woodbury's experiments in this area survive, while only very few examples were published *(plates 18, 52 and 53)*.[134] Ironically, Woodbury's pioneering effort to deconstruct the photographic image led in some part to the development of the halftone, which superceded the Woodburytype.

Woodbury made further adaptations to the process for non-photographic applications. One such incarnation he called photo-filigrane.[135] In this process a piece of paper was run through a roller press with a Woodburytype matrix *without* ink, producing a mock water-mark *(plates 54a and 54b)*. It was intended as a cheap substitute for porcelain miniatures as well as a tool for paper manufacturers. Woodbury hoped the process would be adopted by the paper industry as an alternative to the traditional methods of water-marking. He also proposed using the photo-filigrane as an anti-counterfeit device for printing money, but it was discovered that the images disappeared when the paper got wet; the fibers returned to their original condition. Even though the well-established portrait firm Brown, Barnes & Bell bought the rights from Woodbury, no published prints have surfaced.[136]

Another example is the adaptation of the Woodburytype for decorative wood-burning, where the matrix was heated to burning temperature and pressed onto a piece of wood, or other material, branding the design.[137]

The major problem with these modifications was the dependence on the early steps of the Woodburytype process. It was cumbersome and time consuming when simpler methods like photogravure and halftone achieved the same result.

The only modification that seems to have produced any substantial results was not a change in technology or direction, but a simplification of the process itself. Woodbury knew that his process was suitable only for a publisher or other high volume distributor. With the expense of unique and new equipment required to produce a Woodburytype, large production runs were necessary to cover the overhead. It was an unsuitable process for the hobbyist or artist, who typically wanted only a few prints from any negative. But since the majority of activity in photography at the time was still amateur, Woodbury wanted to devise a process that was more accessible.

He found that he could use the gelatin relief itself as a printing plate, circumventing the need for the hydraulic press. In an early stage in the development of the process Woodbury tried to electrotype the relief, but this was never successful. Using the same principle, he discovered that if the relief were made with a positive instead of a negative and faced with tinfoil it would act just like the lead plates. Named from the Latin *stannum* for tin, the Stannotype was the home version of the Woodburytype and was almost identical, except the matrix didn't have to come from a hydraulic press *(plates 55a-57)*.[138] Anyone with a basic darkroom and knowledge of photographic chemistry could produce Woodburytype prints. Even though the photographic journals published numerous articles written by Woodbury and others, confirming its popularity, the large firms never used it in quantity because the prints still needed to be tipped-in for publication.

By the mid 1890s, the halftone process had become the most common method for reproducing photographs. It is based in part on Talbot's screen principle where the image is broken into small dots that hold areas of solid ink. It can be locked up with type in a printing press and treated like a traditional woodblock plate, or adapted to virtually any other printing technology.[139] Finally a homogenous page of text and illustration could be printed at one time. It could also be used to print in color. Newspapers and other cheap circulars could easily reproduce facsimiles of photographs using this technology, guaranteeing its commercial acceptance.[140] It eventually superceded all previous photomechanical printing processes, including the Woodburytype, and is the basis of all modern, commercial reproductive printing.

Although the halftone doesn't possess the detail and subtleties of the Woodburytype (or other photomechanical processes likes the collotype), quantity and economy were more important than quality. Readers of newspapers and circulars just weren't looking for the kind of high quality image offered by publications like the art and science journals of the time.

In an 1890 listing of Goupil's processes, which had been taken over by Boussod, Valadon & Company, there is no mention of the Woodburytype, indicating they may have started to phase out this production line.[141] In the 1894 Photo-Mechanical Prints Exhibition in New York, Goupil displayed several Woodburytypes, but whether these were new products or overstock, or just a historical display is uncertain. By 1900, gravure reproductions of paintings were the majority of their output.

Limited edition books and journals continued to use hand pulled prints well into the twentieth century. But most firms favored the collotype process for small editions as it could be run on a traditional printmaker's press.[142] This process, brought to perfection by Josef Albert in 1868, was also based on a bichromated gelatin relief, but was printed like a lithographic plate. It didn't have the true half-tones that the Woodburytype possessed, but the grain structure was refined enough that it became the leading competitor to the Woodburytype and eventually succeeded it for fine printing. The cost of printing a collotype was extraordinarily small, a full shop could be set up for about one fifth the cost of a Woodburytype facility.[143] With a large enough run, a single print could be had for about one shilling (1870). But even for the increased economy, Waterhouse wrote, "in beauty, transparency, and delicacy of gradation, the Woodbury-type prints are undoubtedly superior to collotypes."[144]

In 1870 H. Baden Pritchard, who had written several articles on the Woodburytype's operation, compared the differences of the collotype and Woodburytype. In his closing remarks he said:

> *The two processes appear to us capable of different applications, and not liable to clash in any way with one another. On the side of the Lichtdruck [collotype] is the element of cheapness and the feasibility of producing copies of any dimensions, qualifications which cannot fail to secure extensive employment for it in certain channels; Woodburytype prints on*

6 Evolution of the Woodburytype Process & Woodbury's Later Years

[Photography has] revolutionized the engraver's art . . . which cannot fail to have a most important effect upon the illustrations of our literature, and indeed upon pictorial art generally, inasmuch as we no longer depend upon line engraving, woodcutting, or lithography, nature herself reproducing her own drawings at a cost infinitely less than we have hitherto paid for inferior productions of the human hand.

—Andrew Wynter, *Peeps into the Human Hive*, 1874

On March 17, 1865, the *British Journal of Photography* published an article titled "Woodbury's New Printing Process."[145] Woodbury's name was fairly well known from his photographs of Java and several articles describing his adventures on the island, but he was by no means known as an inventor. Woodbury wrote the article in typically vague Victorian language and only gave an outline, not details, of a process for making photographic reliefs. He apparently sent a few sample prints along with his article and said they were "entirely different from any press photographs that have hitherto appeared," and that their chief virtue was the print's grainlessness and identity with a regular silver print. Even though he admitted the process hadn't been entirely worked out, he claimed that a proficient operator could turn out 200 prints an hour operating three presses. The editors attached a brief note to the end of the article:

> *We look forward with much interest to the working details of Mr. Woodbury's extremely promising process. The specimens he has forwarded to us cannot be distinguished from ordinary silver or carbon prints.*[146]

But it took the photographic community some time to realize the merits of the process. Among the hundreds of other photographic developments being described in the photographic journals each year, it may have gone unnoticed had it not been for the immediate controversy over its origins. Another photographic inventor, Joseph Wilson Swan, well known for his developments with the carbon process, had shown prints nearly identical to Woodbury's to the London Photographic Society just a few months previously. Swan and Woodbury knew of each other's developments and had at one point agreed to work together. But the partnership did not last and when Woodbury made his initial announcement, the fight for primacy entered the photographic press.[147]

Both inventors were drawing from a rich tradition of photographic research. The advances Vauquelin, Ponton, Talbot and others made paved the way for the Woodburytype and for carbon printing. In his writings on the collotype, Poitevin described a method of including the pigment directly into the bichromated gelatin and washing the exposed film with warm water to remove the insoluble portions, leaving an image dark in the thick parts, yet light or clear in the thin parts. This is the basic carbon print.

Early experiments with carbon printing suffered one major flaw—as the light strikes the bichromate tissue in proportion to the density of the negative, it only penetrates part way in the mid values.[148] The insoluble part of the tissue then remains on the surface with a soluble portion underneath. When developed in hot water, these half-tones washed away because they were not attached to the base. For more than twenty years, printers tried to solve this problem, using various approaches like exposing from the back of the tissue.

Swan's major contribution to carbon printing was the transfer method. After exposure, the gelatin film is immersed in water, removed entirely from its original support and transferred to another piece of paper on the reverse side. This simple act switched the front and back surfaces of the tissue and the lower soluble portions could now be reached from the front, leaving the half-tones attached to the base. The carbon transfer was the first practical carbon process and became extremely popular with amateurs and professionals, which brought a degree of notoriety to Swan. The press even referred to carbon printing as Mr. Swan's Process *(plate 58)*.

Woodbury says he used a primitive carbon process while in Java to print his photographs and, upon observing the relief nature of the finished prints, speculated that this might be used to make castings.[149] Poitevin had used similar reliefs to make plaster casts and also to form electro-type molds.[150] These earlier processes developed into surface printing techniques using ordinary printer's ink. The idea to use it to make a casting in the very substance of which carbon prints are made—pigmented gelatin, producing a true carbon print by mechanical means—brought about an avenue for permanent, truly continuous tone prints without the use of light or chemical baths.

The regular tissue used for carbon printing, however, was too thin to produce a sufficient relief in the mold and Woodbury discovered he would need a thicker tissue, capable of holding more detail. Once he produced a tissue with enough depth, he could make a mold, but a plaster cast, like the one Poitevin used, could not withstand the force of a press. So Woodbury made an electro-type mold by coating the gelatin relief with a metallic dust and plating that with copper. Once the deposit was thick enough to withstand the force of a press, the original tissue was removed and the electro-type exhibited a negative facsimile of the image. He poured the pigmented gelatin "ink" into the electro-type mold, and after the gelatin set, it remained durable, waterproof and lightfast. The prints produced were identical in almost every aspect to a traditional carbon print.

Toward the end of Woodbury's first article, he mentioned that he had shown a print to a friend of Mr. Swan, and:

> *He acquainted me with the fact that that gentleman [Swan] had also had a similar idea to myself; but, as he learned that I had patented the process, he did not proceed with it. Still, as the idea of the thing occurred to him, independently of any information from me, credit is equally due to that gentleman.*[151]

In a letter to the *British Journal of Photography* dated April 7, 1865, Swan wrote:

> *Will you have the goodness to allow me to correct an inaccuracy in the communication of Mr. Woodbury . . . the friend referred to [by Woodbury*

in his announcement on March 17, 1865 in British Journal of Photography] gave me to understand that this process was not included in Mr. Woodbury's patent. The idea of this new system of printing originated with myself now more than a year ago, and was first made public in an article in the Photographic News *of October 14, 1864.*[152]

Swan had been experimenting with relief carbon prints for some time and in an article in February 1864 on his carbon process Swan wrote, "I have discovered the means of . . . photographic engraving for *intaglio printing*, for photographs with no lines, but only gradation of colour, or light and shade."[153] He said he had been speculating about this process for twenty years. In a subsequent letter to the *British Journal*, Swan gave the specifics of the *Photographic News* article, with the prefatory statement:

When I was in London a year ago to read my paper "On Carbon Printing" before the Photographic Society, I communicated to Mr. Simpson, the general principles and some of the details of a new system of printing for producing by mechanical means pictures almost identical with my carbon prints. . . . I asked him not to publish my ideas until I should have leisure to develop the process further.[154]

George Wharton Simpson was an active member of the photographic societies as well as editor of the *Photographic News*, to whom both Swan and Woodbury frequently wrote letters. Simpson became the key intermediary between the two disputing inventors. In a letter to Swan, Simpson wrote:

Mr. Walter Woodbury, of Manchester, has been experimenting for some time with chromated gelatine, and has just patented a method of getting china transparencies in relief . . . and the electro. from which it was produced, the image being obtained in gelatine and colour. He did not seem to attach much importance to it, and I don't fancy it is included in his specifications.

The article Swan referred to, titled "A New System of Printing," was published on October 14, 1864 by the *Photographic News*. This article gave all the working details of Swan's process, which he had only hinted at before. In it the editors state:

> *We are about to bring under the attention of our readers, a new system of printing, different in principle and detail from any of which we know . . . the originator of the system we are about to describe is Mr. J.W. Swan, whose carbon process is now before the public.*

The method described is exactly the same as what Woodbury published in March 1865. The article about Swan continues:

> *A few weeks ago, however, Mr. Walter Woodbury, of Manchester, called upon us, for the purpose of showing us examples of his experiments in producing photographs in relievo and intaglio . . . he produced a rough print obtained by the very method which Mr. Swan had described to us some months previously. Thus the idea had occurred to two gentlemen independently. Mr. Woodbury did not, however, regard the principle as capable of very successful application, whilst Mr. Swan regards it as of the utmost importance, and capable of the most valuable development. . . . The print by Mr. Woodbury, although a rough experiment, was to us a very interesting and suggestive result; but Mr. Swan has made greater progress.*

Woodbury wrote in his March 1865 article that, "although in the specimens sent [sic] is not so perfect as might be wished, still promises the greatest success."

To summarize this chain of events: Swan spoke of his process in February and again in April 1864, Woodbury received his patent that September, Swan published full working details in October and Woodbury did so in March 1865. Why Woodbury waited so long to publicly announce his developments is unclear. It is possible that, upon hearing Swan was working on the same ideas, he rushed his patent through even though there were still problems with the working method.

According to Swan, after hearing about each other's developments Woodbury suggested they should form a partnership.[155] The two met on February 28, 1865,

Leonardo was printing directly from the leaf itself. Even though this produced prints with extreme fidelity, the leaf could not withstand more than a few impressions before it broke down. Later experimentation gave rise to the lead plate impression and even later to electro-types taken from the lead plates. In whatever incarnation it was used, prints were made that duplicated the fine details of the specimen without any manual re-interpretation.

The re-interpretation of an artist's work was Wallis's main concern, for he was an artist himself, dissatisfied with having his pictures re-drawn for the press. He said he developed the process for want of a way to reproduce his own drawings "without the intervention or translation of another mind and hand, as was now necessary in engraving."[164]

Because Wallis was not a photographer, he gave primary importance to executing drawings. He did recognize the potential of the process, however, and described a method for obtaining prints from photographs. The *Photographic News* immediately reprinted Wallis's paper "The New Art of Auto-Typography," published only days before in the *Society of Arts Journal*.[165] He also attended the meeting of the South London Photographic Society in May 1863 to discuss the possibilities the process held for photography.

At first Wallis was content with using photographs as a copying aid, but later experimented with sheets of bichromated gelatin upon which an image could be exposed, producing a relief, just like the carbon process. Wallis was capitalizing on the newly discovered property of gelatin to possess extreme hardness when dry. He found it was hard enough to withstand great pressure and will impart every detail when pressed into a sheet of lead. He combined his drawing method with Poitevin's ideas on bichromated gelatin film to produce photographic plates.

Then, a full year had passed without the slightest mention of Wallis when an article in the *Photographic News* proposed the idea of combining Wallis's plate making process with the new developments Swan had made with the carbon transfer process.[166] If this could be done, the editors speculated, it would produce mechanical photographic prints with true half-tones.[167] Swan had evidently started experimenting toward this end as he had sent sample prints (although it is unclear when) pulled from metal plates, which the editor said showed "a considerable amount of half-tone."

> *The question now arises, whether the ingredients necessary to give a sufficiently hard surface, used in Mr. Wallis's ink, might not be incorporated with gelatine, instead of the pigment used in Mr. Swan's process . . . we see no reason why at least an approximation to a photographic half-tone might not be also obtained by using photographs produced by such a modification of Mr. Swan's process as we have indicated.*

Why did Wallis's process, identical in conception to both Swan's and Woodbury's, never gain any real success? Despite all of his efforts and the support from the photographic community, he could not get a photographic image of satisfactory clarity. Analysis of the specifications reveals a few flaws, which Wallis, concerned mostly with drawings, never worked to overcome. Wallis would have run into the same problem noted by early experimenters with the carbon process: when a sheet of bichromated gelatin is exposed from the front, the middle values wash away, resulting in a high contrast image unsuitable to printing photographs.[168] His technique would best be suited to images like drawings. After the carbon transfer process, we don't hear any more from Wallis about this. Another problem—impressing the relief into a sheet of lead by running it through a roller press, as Wallis describes—would produce uneven plates varying in thickness. All the irregularities in the matrix would make it virtually impossible to pull prints of any quality.[169]

Through Woodbury's experiments, both of these problems were solved. He overcame the exposure problem by experimenting with a number of clear substances, onto which the gelatin would be coated, eventually settling on collodion.[170] Once dried, this film could be peeled off its support, the gelatin exposed from the rear and re-attached to a new base before development, somewhat like the carbon transfer process. By 1864, neither Swan nor Woodbury had hit upon the idea of using lead plates. At first both were using electro-type molds, but Woodbury experienced problems with them and continued to experiment. One of these problems was the gelatin had the tendency to swell when immersed in the electro-type fluid, which caused the image to lose detail.[171]

Though later writers spoke of nature printing as the ancestor of the Woodburytype, Woodbury himself never gave an indication of where he got the idea to use lead plates. It was not covered in his original patent and the first time he mentioned it was August 1865. He said one could use either a roller press (like Wallis had done) or a hydraulic press.[172] But Woodbury settled on the flat-bed because it was more accurate, since all areas of the plate received an equal amount of pressure at the same time.[173]

The lead printing plate was the last major development photo-relief printing needed to make it commercially successful. It was in this form that it was practiced for most of its life and since Woodbury perfected its use, the process became known as his. The need for a lead plate also kept the process out of the reach of amateurs. Only a press of extraordinary pressure could produce these lead plates and they were beyond the means of all but the largest commercial firms.

There are no records that Swan worked to overcome the problems with electrotype plates and this may have been one of the reasons he stopped working on the process. On July 6, 1865 he finally received a patent for his "Improvements in the production of printing surfaces by photographic agency, and in obtaining prints therefrom."[174] He called his method *Photo-Mezzotint.* The process was so close in content to Woodbury's method and used exactly the same materials that later writers referred to them as the same.

◆

Unlike Daguerre who gave his invention to the people—for a handsome stipend from the government—William Henry Fox Talbot patented his negative-positive process. This initiated a model in the British photographic world where commerce came to play a more prominent role than art. Woodbury's patents made it possible for him to keep a close rein on the few people licensed to use it; he even revoked privileges from one firm that had not kept up payments.[175] But this plan backfired on him when, for unknown reasons, the patent expired in 1872 and the process passed into public domain.[176]

Since the Woodbury Permanent Photographic Printing Company, of which Woodbury had no stake, had acquired the rights to Swan's photo-mezzotint process they were able to maintain some legal control over photo-relief printing. The firm continued to use the moniker *Woodburytype*, effectively ending the fight for primacy.[177]

Other firms took up the process, but the WPPPC still produced the majority of prints. This seemingly unfortunate act may have been the best thing for Woodbury as it forced him to reconsider his stability. He started working on other experiments that led to the development of the Stannotype, around 1880, which he also patented.

By this time other processes had captured the attention of the photographic community. But as soon as Woodbury released details about the Stannotype, affinity for the old process returned. It gave amateurs an easy way to produce Woodburytype prints of their own work and it gave professionals the chance to use a formerly controlled process. The photographic press was inundated by people who had suggestions for improvements, hints on technique and formulae for the process. All the photographic manuals and journals of time published thorough working details.[178] A number of firms licensed the Stannotype from Woodbury (see appendix); however, since the prints are indistinguishable from the earlier Woodburytypes, it is hard to say if any actually were published.[179]

Despite the problems early in its life, the Woodburytype became the most widely used method of its time for reproducing photographs and other art for publication. In 1881, *Anthony's Photographic Bulletin* wrote, "[The] Woodburytype . . . is and apparently will be for many years to come, the only photo-engraving process of practical and commercial value."[180]

In the last quarter of the nineteenth century, however, changes were occurring in the photographic community. The advent of the dry plate and soon thereafter of roll film, began the rise of mass-produced photographic goods. Photographers no longer needed to make the majority of their own materials. All these changes in technology brought about the eventual decline of the commercial use of the Woodburytype. In an effort to stave off the inevitable loss of revenue, the Woodbury Permanent Photographic Printing Company started manufacturing non-Woodburytype materials such as carbon tissue and dry plates, but to no avail.[181]

The photographic community widely recognized Woodbury's inventive genius during his own lifetime. "In a work by a French writer, Davanne, this name is placed next in rotation to those of Talbot, Daguerre, and Niepce."[182] He sat on numerous photographic committees in England and on the continent and was elected to the council of the British Photographic Society.[183] The French Photographic Society awarded him a special medal in 1871, the Photographic Society of Belgium awarded him a gold medal for the Woodburytype in 1876 and the English Photographic Society awarded him a progress medal for the Stannotype in 1884.[184]

These honors were a sore subject for Woodbury, however, because by this time he was already having business difficulties and he felt real recognition should have come in the form of financial support. He declined the seat on the photographic council, weakly claiming he lived too far from London. Similarly with a medal:

> *I may allude to a rumour I have heard; it is to the effect that Mr. Woodbury has declined to accept the medal which was offered to him. How could he act otherwise? After receiving special gold and silver medals from the foreign Societies more ready to recognize merit, it was not to be supposed that he would accept at second-hand a bronze medal from the London Photographic Society, by whom it was only offered because they had nothing else to do with it.*[185]

Woodbury continually experimented with new ideas throughout his life making advances in optical lanterns, stereoscopes, kaleidoscopes, barometers, hygrometers and was a pioneer in balloon photography.[186] He patented these devices too, and wrote articles detailing his experiments.[187] One report claimed Woodbury had taken out thirty patents.

Despite Woodbury's acclaim as an inventor, the expiration of his patent rights, the failure of his businesses and the continued controversy over the real originator of the Woodburytype made his life constantly unstable.[188] Neither did he ever achieve substantial success as an artist. Relatively few of his images were ever published, mostly because he became so caught up in the business of the Woodburytype.[189]

In the early part of 1885, shortly after the failure of another one of his businesses, Woodbury's health again began to fail. The photographic community established a fund for his recovery that ultimately raised £316 15s, "a considerable portion of which was expended carrying Mr. Woodbury over a period of three months' serious illness."[190] It appears that Woodbury kept the true cause of his illness a secret and it was only revealed after his death by Edward Wilson, a long time friend and editor of the *Philadelphia Photographer*. In his last letter to Wilson, Woodbury wrote, "I had a bad time of it with 'diabetes,'—a nasty, weakening disease—but under proper treatment my health has much improved."[191]

He recovered for a short time before he died at Margate, England, during the night and was found Saturday morning, September 5, 1885.[192] He was only 51 years old. The *British Journal of Photography* reported:

> *For some time past the deceased had been compelled to resort to the use of opiates for the purpose of procuring sleep, and on this occasion he took his usual dose, telling the children he "was very tired and wanted a good sleep, and that if they did not make a noise and wake him too early in the morning he would take them for a donkey ride." Accordingly the children next morning went quietly out, and on returning between eleven and twelve o'clock found him still in bed, and could not awake him. Upon calling the landlady he was found to be dead.*[193]

An inquest was called to look into the matter and the doctor determined Woodbury died from an overdose of laudanum, assuming he had used it as a sleeping aid. The *Photographic News* reported that, "to a man of Mr. Woodbury's habits and temperament, after the fatigue of the journey [to Margate], the usual dose might possibly be too much."[194] Someone implied it was suicide, but no evidence surfaced to support that claim.[195] His funeral took place at Abney Park Cemetery, Stoke Newington, London *(plates 60a and 60b)*.

Wilson wrote an editorial in the *Philadelphia Photographer* of quite bitter sentiment. He considered Woodbury *(fig. 11)* the finest photographic inventor ever and his process the most important and still capable of "great expansion."[196]

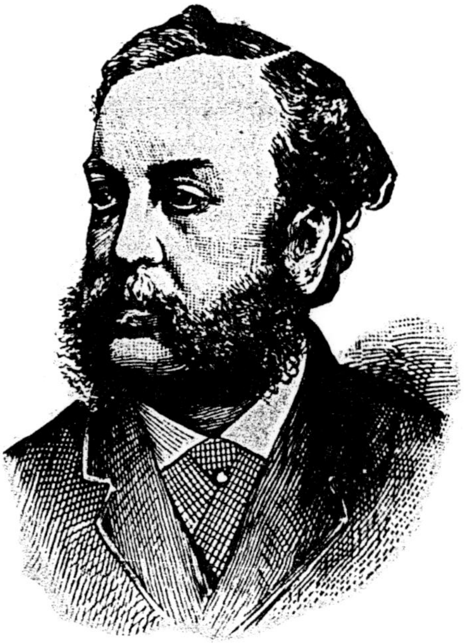

Figure 11. Portrait of Walter B. Woodbury. Block print published at the time of his death in the Philadelphia Photographer, *September 1885.*

> *Walter Bentley Woodbury is dead. The sad news comes to us through our English exchanges, just as we go to press. The warm-hearted friend, the able scientist, the ingenious inventor is dead. To us it is a personal bereavement, for we had been friends for nearly twenty years. . . . We do not hesitate to place him at the head of all the inventors in our art, who ever lived. He was only fifty-one years old, and it will be fifty more before our loss can be repaired.*[197]

In a following issue Wilson continued his emotional outpour in tribute to his dead friend and said,

> *Lovely, amiable, tender-hearted, generous Woodbury, how our heart sank when we learned of his premature end, and how our fists clenched and resentful feelings arose. Some have dared to insinuate that our friend was a suicide. Shame upon them! He was murdered rather. There are men—wealthy men—alive now upon whose skirts his blood-stains are. Men who have robbed him of his rights—who are making money, and have been doing so for many years, from the fruits of his inventive genius, and upon whom we call now to divide with the widow and the fatherless, before their names are made public.*[198]

Woodbury was survived by his wife and eight children, seven girls and one son. Walter Edward Woodbury, born on his father's 31st birthday, June 22, 1865, followed his footsteps, becoming deeply involved in the photographic community. He was editor of the *Photographic Times* and other smaller photographic journals, and author of several books including *The Encyclopaedic Dictionary of Photography*, which gives the most complete technical description of the Woodburytype process in the English language.[199] Walter E. Woodbury died on June 26, 1905 in Panama.[200]

Josef Maria Eder wrote in his *History of Photography* that the last Woodburytype press was sold for scrap iron in 1928 after faster processes had displaced it. But he lamented its disappearance.

> *Woodbury's process disappeared entirely from the graphic field in the nineties, displaced by heliogravure, the halftone process, Klic's rotogravure, and intaglio rapid-printing presses. These processes are capable of producing large editions in a very short period of time, and require no cutting and pasting of individual pictures on pasteboard; they permit the incorporation of illustrations in the text and allow the execution of large sizes; but they have never been able to attain the soft gradations, the superb rendering of middle tones and modulated shadows characteristic of the Woodburytype.*[201]

◆◆

AFTERWORD

While writing this book, several things about the Woodburytype process and its history became clear to me. Although I did have to include some personal conclusions in the main text for clarity, I decided to put most of them in the form of an afterword.

— Art vs. Commerce —

Born at the height of the industrial revolution, it seems as though photography as an art has always taken a backseat to commerce. Most developments in its history have been to make it easier, faster and more portable. Even today we are witnessing a dying off of superior products, and a plethora of simpler, yet not quite as refined materials appear daily. A paper like Agfa's graded Portriga, long the best warm-toned paper around, has ceased production in favor of a variable contrast paper that just doesn't have the same depth. Kodak's Super Double-X film, a staple for many artists, has disappeared for lack of demand. And just as this book is being finished, Kodak has announced that it is stopping all black and white paper production. Many of the best products are no longer available and the selection for serious photographers is becoming severely limited. The aesthetic of photography is now being dictated by the manufacturers. I believe this is one of the reasons behind the recent revival of the nineteenth century processes. In many ways these older, manual crafts restore artistic control to the photographer.

Because of this perpetual trend toward convenience in manufacture, the Woodburytype was doomed from the beginning as a commercial process. Even though it offered a tremendous advance over existing reproduction methods, it became outmoded before it reached its full potential. I believe a few things kept it from staying in use longer.

The initial outlay and additional labor required to print Woodburytypes made it a bad investment with the ever increasing demands for volume. The process could not keep up with text printing. In essence it was a craft process at the beginning of the industrialization of photography—a kind of evolutionary link between the traditional and the fully mechanical print. It was oddly and ideally suited to the mass production levels of 1870, but was no longer applicable by 1890. That it gained any acceptance and use is evidence of its aesthetic superiority. Had Woodbury realized the essence of the process was its inherent beauty and fidelity to the original and had he presented it in this light, as a niche product for richly produced editions, he might have been able to keep it alive.

Even though the first published Woodburytype was an art photograph by Robinson and many of the later reproductions were paintings, the process never really caught on as an artist's medium. The need for a hydraulic press—not solved until about 1880—kept it out of the reach of individuals who wanted only a few prints. But by the time this was solved, the Woodburytype was pigeonholed in the mass production market and other processes like the platinum print and photogravure had caught the attention of artists. Had the Woodburytype been introduced, in the form of the Stannotype, forty years later, during the height of pictorialism and the Photo-Secession, it might have been more appreciated as an art medium. But no one in the U.S. was printing Woodburytypes by the time Stieglitz started publishing *Camera Work*, which relied heavily on craft-based reproduction methods for its illustrations. The Woodburytype could have been another tool in their arsenal. And even though it does not have the romantic softness we associate with this movement, it would have captured all the subtleties of handwork and manipulation in these artists' photographs.

Additionally, I believe Woodbury tried to maintain too much control over the patent rights. A superior technology is sometimes lost because of too strict patent controls, which leads to limited application and, like a family line with no progeny, dies off within a single generation. As the collotype and photogravure continued to be used for niche printing, so would have the Woodburytype, except a small shop could not afford the cost of a patent license. He may even have made his profit by producing materials for the process, teaching it to others and continuing to run a shop with high volume and the highest quality output. He appears, however, to have been more interested in the commercial success attained with mass production.

Had Woodbury given more public access to the process and let numerous people use it without license, it would have lasted longer. Printers would have inevitably modified it to suit their own purposes, adapted it to more modern technologies and it would have mutated into a family of processes, each appropriately suited to a particular demand. From all the articles in the photographic press written by people other than Woodbury, it is obvious the process had a life of its own.

— Inventor —

The controversy between Woodbury and Swan over the originator of the process occupied a great deal of my research attentions. In a way, it is almost irrelevant who had the idea first and who made the first prints. We have the process now. But I do not wish to perpetuate a myth and want to give credit where it is due. The parts Swan and Wallis played in shaping the process must not be overlooked. Woodbury did not invent photo-relief printing. What he did accomplish was the perfection of the Woodburytype. As Swan said, Woodbury should be given full credit for bringing the idea into commercial application. Indeed without Woodbury and the controversy, the process may have gone unnoticed and unused. But in the course of my research, I was overwhelmed by the evidence in favor of Swan's primacy on the invention. He was the first to show samples to anyone of importance in the photographic community, the first to have published an article detailing the process, and the previous fact of his perfecting the carbon process weighs heavily in his favor.

— Today —

Unlike previous graphic media, which enjoy a long tradition of intellectual appreciation, photography is still finding its way. With the enormous amount of scholarship in the last few decades and a growing appreciation of process as an integral part of the art, photography can now truly stand on its own. In the future we will be acquainted with many artists, both new and old, not in the long canon of museum exhibitions. And we will come to understand many of the forgotten processes that have played such a vital role in shaping the photography of today. With some of these processes enjoying a revival, it has been proven that no medium becomes obsolete, only that people don't know what to do with it. The Arts & Crafts ideal of truth to material extends to process. But it takes dedication to discover potential.

Unrivaled in grainlessness, ability to resolve detail, color and tonal range and its resemblance to a real photograph, the Woodburytype stands as one of the great achievements in photographic technology. It is extremely adaptable to the craftsman, giving unlimited possibilities for artistic statement and it deserves to be part of the history and language of today's photography world.

◆

After adjournment, some fine Woodbury transparencies, made by the American Photo-Relief Printing Company, were shown in the lantern.
—Philadelphia Photographer, 1872

x THE PHOTOGRAPHIC NEWS. [MAY 9, 1884.

STANNOTYPE. { TO FOREIGN AND COLONIAL PHOTOGRAPHERS.

In consequence of numerous inquiries from abroad respecting the above Process, Mr. Woodbury begs to inform those interested that SOLE RIGHTS for the following Countries have been already disposed of :—

GREAT BRITAIN and IRELAND—Messrs. WOODBURY, TREADAWAY, & Co., 35, Bucklersbury, E.C. FRANCE—M. HUTINET, Avenue Parmentier, Paris. GERMANY—Messrs. PRUMM & SCHAARWACHTER, Unter den Linden, Berlin. BELGIUM—Mr. T. J. BULL, Rue Royale, Brussels. AMERICA—Mr. F. N. CLARKE, Rood Lane, E.C.

Mr. WOODBURY is still open to treat for SOLE RIGHTS to work the Process in the following Countries & Colonies :—

SPAIN, PORTUGAL, ITALY, AUSTRIA, RUSSIA, HOLLAND, SWITZERLAND, DENMARK, SWEDEN & NORWAY, CANADA, VICTORIA, NEW SOUTH WALES, SOUTH AUSTRALIA, NEW ZEALAND, TASMANIA, CAPE, INDIA, CEYLON, BRAZILS, MEXICO.—For terms apply to—

WALTER B. WOODBURY, Java House, South Norwood, London, S.E.

Figure 14. Advertisement soliciting patent rights inquiries for the Stannotype process published in Photographic News, *May 9, 1884.*

GOLD MEDAL, INTERNATIONAL INVENTIONS EXHIBITION.

SILVER AND BRONZE MEDALS, LEEDS EXHIBITION.

WATERLOW & SONS,

LIMITED,

Photographic Art Printers and Engravers,

FINSBURY WORKS, LONDON, E.C.

Photo-Zincography.

Photo-Lithography.

Photo-Mechanical Printing.

Collotype.—Hand and Machine Printing for Art Publications, Scientific and Antiquarian Periodicals, Machinery, Landscapes, Portraits, Pottery, Furniture Designs, Trade Advertisements, &c. The superior results given by this process, and the rapidity and cheapness by which the prints are produced, together with the advantage of printing with or without margins, place it in the first rank of processes for commercial purposes.

MESSRS. WATERLOW & SONS, LIMITED, have given this branch of the Photo-Printing Department special facilities for the production of good work, and have introduced the most perfect machinery and plant obtainable.

Woodbury Prints.—High-class permanent copies, equal in appearance to the best Silver Prints, of Portraits, Landscapes, Furniture, Pottery, &c. Prints may be obtained in almost any colour from Customers' own Negatives, or from the original objects. These reproductions are specially suitable for Portrait work, and are valuable for every description of Artistic or Commercial Illustrations.

Photo-Zinco Engraving.—Blocks for Surface Printing, from Line and Grained-paper Drawings, Steel and Copper Plates, Wood Engravings, &c., &c. Letter-press Blocks in "Half-tint" (stipple or dot) direct from Photographs from Nature, without drawing.

Accurately registered Blocks for Chromographic Printing. Intaglio Engraving in Line and Half-tone on Copper and Zinc.

The greatest care and skill is employed in the production of these Blocks, and the results are the finest which it is possible to obtain.

The new and extensive Photographic Works, being fitted with Modern Appliances, Machinery, Electric Lighting, &c., rapid and accurate work is always obtainable, irrespective of weather or season.

PRICE LISTS, ESTIMATES, and full particulars on application.

WATERLOW & SONS, Limited, Finsbury Works, E.C.

Figure 15. Advertisement for Waterlow & Sons published in Photography: Its History, Processes, Apparatus and Materials, *by A. Brothers, 1892.*

THE

WOODBURY PERMANENT PHOTOGRAPHIC PRINTING COMPANY,

EYRE & SPOTTISWOODE, PROPRIETORS.

BEG to call the attention of Publishers to the advantages of the **Woodbury-Gravure** process for the purpose of High-class Book Illustrations, it being a process which reproduces with equal facility copies of wash or line drawings, portraits from life, landscapes, photographs, paintings, engravings, and sketches of all kinds. **Woodbury-Gravure** approaches nearest to photo-engraving in results, but without the initial expense of preparing the plate.

The WOODBURY COMPANY are also the original holders of the patent for Woodbury type printing, and which they have most successfully worked for many years, the results produced being far superior to any other mechanical process for the reproduction for either Book Illustrations, Trade Catalogues, &c., &c., where mounting is not an objection.

Examples of both the above-mentioned processes will be found in this Publication.

The WOODBURY COMPANY also undertake

Collotype Printing . . for Book Illustrations.
Carbon Printing . . for Publication Purposes.
Platinotype Printing . for all Requirements.
Silver Printing . . . of the Best Description.
Carbon Enlargements on the Company's own special make of Tissue from Amateur or Professional negatives.
Carbon Tissue . . . in Cut Pieces for small Consumers.

PRICE LISTS and all Information on application to

6 GREAT NEW STREET,

LONDON, E.C.

WORKS—CASTLE BAR, EALING, W.

Figure 16. Advertisement for The Woodbury Permanent Photographic Printing Company published in Photography: Its History, Processes, Apparatus and Materials, *by A. Brothers, 1892.*

THE WOODBURY PERMANENT PHOTOGRAPHIC PRINTING COMPY.

EYRE & SPOTTISWOODE (PROPRIETORS),

6 GREAT NEW STREET, LONDON, E.C.

WORKS: CASTLE BAR, EALING, W.

DEPARTMENTS FOR

WOODBURY-GRAVURE (Patented). This new adaptation of the Woodbury process is most effective, and entirely obviates mounting and cockling. It approaches nearest to Photo Engraving for Fine Art Book Illustration.

WOODBURYTYPE PRINTING for Trade Catalogues of all kinds—Portraits, Landscapes, Articles of Manufacture, &c.

COLLOTYPE PRINTING. This process has been much improved. Prices are the lowest consistent with good work.

CARBON PRINTING for Publication purposes. Our numerous publications in Carbon are a proof of the utility of this process.

CARBON ENLARGEMENTS. We have several unique lines in these, at prices which compare very favourably with Bromide.

CARBON TISSUE. Supplied in convenient quantities for Amateurs' and small consumers' use, either sensitive or unsensitive. ¼-plate, 5d.; ½-plate, 10d.; 1/1-plate, 1/6; 10×8, 2/-; 12×10, 3/-; 15×12, 4/- per dozen. Posted free to United Kingdom.

PLATINOTYPE PRINTING for all requirements, and in any quantity, for trade, publication, or Amateurs.

SILVER PRINTING of the best description, in large or small quantities.

AMATEURS' OWN NEGATIVES Developed, Varnished, Spotted, Blocked out, Printed from, Mounted, or Enlarged.

INSTRUCTION IN CARBON PRINTING FOR AMATEURS. A concise Pamphlet, with Specimen Print. Post free, 4 stamps.

Price Lists, Specimens, and all Information on application to

THE WOODBURY PERMANENT PHOTOGRAPHIC PRINTING CO.

6 GREAT NEW STREET, LONDON, E.C.

CC

Figure 17. Advertisement for The Woodbury Permanent Photographic Printing Company. Unidentified newspaper clipping, collection of Luanne Woodbury.

- Woodburytypes had a problem with printing large areas of white. This is particularly noticeable in the skies of landscapes. They may be quite mottled or splotchy looking. One of the solutions was to physically wipe off the pigment and sometimes this can be seen by traces left around rooflines and other areas that adjoin the sky.
- Many Woodburytypes frequently display light spots, or "freckles." They may be randomly dispersed, but are more common in areas of thin pigmentation. These are caused by air bubbles during development. While these may also be found in carbon prints, they are much less common.
- Many of the older Woodburytypes display a fine pattern of crazing, while the cracking in carbons is usually much more coarse. This characteristic diminished with time, as the printers were able to use less gelatin in the print. Sometimes when a heavy black vignette border is used, this will display a coarse cracking pattern, but usually will not extend into the image.
- Analysis by x-ray spectrometer reveals carbon prints may contain trace amounts of chromium, a by-product of the potassium bichromate, while Woodburytype prints are less likely to contain this substance (the bichromate being part of the relief, not the print). Woodburytypes also contain trace amounts of lead, a remnant of the lead plate.

— Significant Names Connected with the Woodburytype Process —

This is a partial list of people and companies connected with the Woodburytype process. It is not complete and should be used as a starting point for identifying prints. Some of the companies listed here are known to have published prints, while others were only named by Woodbury as holding patent rights.

American Photo-Relief Printing Company (**APRPC**), 1002 Archer St, Philadelphia. Run by John Carbutt, sold and moved to Boston where it was destroyed by fire.

Robert Howe Ashton, Co-filer with WBW for British patent 947, March 30, 1867.

Boussod, Valadon & Company, See Goupil et Cie.

Adolphe Braun, Dornach, Alsace. Patent holder for Woodburytype.

Vincent Brooks. Owner of Permanent Photographic Printing Company (see Woodbury Permanent Photographic Printing Company).

Brown, Barnes & Bell, 220 & 222 Regent St; 62 Cheapside; 12 Baker St, Portman Square; Liverpool and London. Patent holders for the photo-filigrane process.

Thomas Bolas, London. Friend of Walter Woodbury, writer on photography, donated many of Woodbury's personal items to the Royal Photographic Society.

M. Brukman, Munich. Patent holder for Woodburytypes.

T.J. Bull, Rue Royale, Brussels. Patent holder for Stannotypes.

John Carbutt. See American Photo-Relief Printing Company.

F.N. Clarke, Rood Ln [city unknown] USA. Patent holder for Stannotypes.

George Davies. Co-filer, with Woodbury of British patent 105, January 12, 1866.

A.A.E. Disderi, 70-72 Brooke St, Hanover Square; Hereford Lodge, Old Brompton; 4 Brooke St Hanover Square, London. Started the first Woodburytype company in England.

Eyre & Spottiswoode. See Woodbury Permanent Photographic Printing Company.

Goupil et Cie, Paris (taken over by Boussod, Valadon & Company). Patent holder for Woodburytypes; first company to print Woodburytypes on large scale. Also printed Woodburytypes in colour and Goupil-Gravure.

R. Hoe & Company, USA. Manufactured presses for the Woodburytype.

M. Hutinet, (Hutinet & Lemy) Avenue Perimeter, Paris. Printed Woodburytypes, sold Woodburytype tissue, patent holder for Stannotypes.

Mr. Johnson, Bombay. Patent holder for Woodburytypes.

Lemercier & Company, Paris. printed Woodburytypes.

Louis Ducos Du Hauron. Set up a color Woodburytype press at Ballincourt, but failed due to registration problems.

London Stereoscopic & Photographic Company, London. Printed Woodburytype stereographs.

Negretti & Zambra, London. Publisher of photographs by Walter B. Woodbury and Woodbury & Page.

M. Nelon, Toulouse. Patent holder for Woodburytypes.

M. de Parader, British Columbia. Patent holder for Woodburytypes.

Pierre Patin, Paris. Printed Woodburytypes, patent holder for Stannotypes.

Henry Peach Robinson. First photographer to have an image published in Woodburytype. Advocate of art photography.

M. Pedroni, Bordeaux. Patent holder for Woodburytypes.

Messrs. Prumm & Schaarwachter Unter den Linden, Berlin. Patent holder for Stannotypes.

Sgap, Paris. Printed Woodburytypes.

Simanau & Toovey, Belgium. Patent holder for Stannotypes.

Thomas Sutton. Frequent writer on photographic technique.

Joseph Wilson Swan, Sir. Independent inventor of photo-relief technology. Swan's process was known as photo-mezzotint. Known for his refinements of the carbon process.

Swan Electric Engraving Company, 116 Charing Cross Rd, London.

Tangye Bros. Manufactured presses and hydraulic apparatus for the Woodburytype.

John Thomson. Photographer, friend of Woodbury, co-author with Adolphe Smith of *Street Life in London*, the first major social documentary book with illustrations printed in Woodburytype.

George Wallis. The earliest person known to have developed the ideas on which the Woodburytype process works.

Waterlow & Sons, London. Printed Woodburytypes.

David Winstanley, Jr. Former employee of Woodbury who continued to make refinements on and write about the process.

Walter Bentley Woodbury, Java House, South Norwood, London. Inventor of Process. See also Woodbury Permanent Photographic Printing Company.

Walter Edward Woodbury, 423 Broome St., New York. Walter Bentley Woodbury's son, Author *Encyclopaedic Dictionary* and *Encyclopaedia of Photography*. Both have significant articles on the process. Also sold Woodburytypes and Woodburytype materials.

Woodbury & Page, Java. Photographic Studio owned by Woodbury with John Page and later run by both Albert and Henry Woodbury, Walter's brothers.

Woodbury Gallery, 37 New Bond St, London. Run by Eyre & Spottiswoode ca. 1900.

Woodbury Permanent Photographic Printing Company. Known variously as Woodbury Permanent Printing Co., Permanent Printing Co., Permanent Photographic Printing Co., Woodbury Company, et al. It is still unclear if all these companies are the same or not. Taken over by Eyre & Spottiswoode. 6 Great New St, London, with works at Castle Bar, Ealing; 157 Great Portland St; 9A Herford Square, Brompton; 178 Regent St, London. Woodburytypes, Woodburytype materials, Woodbury-gravure, other photographic processes.

Woodbury, Treadaway & Company, 35 Bucklersbury, London. Formed by Woodbury to print Stannotypes.

E & JB Young & Company, 7 West 18th St, New York. Patent holder for Stannotypes.

BIBLIOGRAPHY

— Annotated Select Bibliography — on the Woodburytype Process & Walter Bentley Woodbury

The following bibliography contains the source material consulted for the main text plus a selection of other works pertinent to the Woodburytype process and Walter Woodbury, including historical references that played a part in the development of the process. They have been arranged in chronological order to show the evolution of the process.

Abbreviations used in the bibliography

BJP—British Journal of Photography
PP—Philadelphia Photographer
PN—Photographic News
PT—Photographic Times
WBW—Walter Bentley Woodbury

1802

Wedgwood, Thomas and Sir Humphrey Davy. "An account of a method of copying Paintings upon Glass. . . . " 1802. Reprinted in *On Photography: A Source Book of Photo History in Facsimile*. Beaumont Newhall, ed. New York: Century House, 1956.

1839

Ponton, Mungo. "Notice of a cheap and simple method of preparing paper for photographic drawing, in which the use of any salt of silver is dispensed with." *Edinburgh New Philosophical Journal* 27:169-71.

1839

Niepce, Joseph Neciphore. "Notice on Heliography." as printed in L. J. M. Daguerre's

Suitable Materials. British Patent 2338 [This is the main patent upon which Woodbury's claims of primacy are based, although, the process outlined in this patent is substantially different than the one that he employed commercially and does not include details for making prints.]

1864, October 14

Swan, Joseph Wilson. "A New System of Printing." *PN* 8:493-4.

1865, March 17

WBW. "Woodbury's New Printing Process" [with editorial comment]. *BJP* 12:134. [This is the first article by WBW on the process and contains all the basic steps with the exception that an electrotype plate is made instead of the latter lead mold. WBW also mentions his awareness of Swan's research and gives him credit.]

1865, April 7

Swan, Joseph Wilson. "Woodbury's New Printing Process." *BJP* 12:182-3. [This correspondence from Swan starts the public debate over primacy of the process. Swan mentions the *PN* article of October 14, 1864.]

1865, April 18

WBW. "The New Photo-Relief Printing Process in Semi-Transparent Ink." *PN* 9:387-8.

1865, April 21

WBW. "Woodbury's Process." *BJP* 12:211. [Woodbury's response to Swan.]

1865, May 5

Swan, Joseph Wilson. "Mr. Swan's Claims versus Mr. Woodbury's." *BJP* 12:238-9. [This letter from Swan details his claims to primacy, giving the circumstances of the earliest dealings with Woodbury, Swan's publication of the process before Woodbury and Swan's claim that Woodbury's patent does not cover all steps of the process. It also includes the specifications of Woodbury's patent, but it differs dramatically from the published patent.]

1865, May 12

"New Patent: Complete specifications of Patent Obtained by Walter Bentley Woodbury, Manchester, filed 23rd March 1865, and entitled——*An Improved Method of Producing or Obtaining by the aid of Photography, Surfaces in 'Relievo' and 'Intaglio' upon Aluminous, Vitreous, Metallic, or Other Suitable materials.*" *BJP* 12:247-8. [The specifications in this patent, including the date of filing, differ dramatically from the original published patent and those in Swan's May 5, *BJP* article. They include details for making prints.]

1865, May 12

WBW. "Mr. Woodbury's claims versus Mr. Swan's." *BJP* 12:254-5. [Woodbury's response to Swan's further claims, with editorial note that the public discussion will be discontinued.]

1865, May 15

WBW. "Woodbury's New Printing Process." *American Journal of Photography and Allied Arts:* 515-6. Reprinted from *BJP*. March 17, 1865. [First notice in American press.]

1865, July 6

Swan, Joseph Wilson. Improvements in the Productions of Printing Surfaces by Photographic Agency, and in Obtaining Prints Therefrom. British Patent 1791, [Called Photo-Mezzotint by Swan, the steps in this patent are identical to Woodbury's process.]

1865, August 18

WBW. "The New Photo-Relief Printing Process in Semi-Transparent Ink." *PN* 9:387-8. [The first article to give a detailed description of and formulas for the process.]

1865, October 13

WBW. "The New Method of Relief Printing." *PN* 9:489-90.

1865, October 20

WBW. "The New Printing Process - A Day With Mr. Woodbury." *BJP* 12.535-6.

1865, December 1

"Woodbury's Photo-Relievo Process." *PN* 9:565 [Announces the forthcoming first published Woodburytype print by H.P. Robinson]

1865, December 15

WBW. "Photo-Relief Printing." *BJP* 12:632. [A demonstration of the process held at a meeting of the London Photographic Society, December 12.]

1866, January 5

"The Past Year." *PN* 10:5.

1866, January 26

"Our Presentation Print - Woodbury's Process." *PN* 10:37-8, 48. [first published Woodburytype print by H.P. Robinson]

1866, February 20

WBW. Mode of Printing Photographs. United States Patent 52,803.

1866, April 27

"Talk in the Studio - The Photo-Relief Printing Process." *PN* 10:203.

1866, May 8

WBW. Improvements in Producing Designs upon Wood and Other Materials by the Aid of Photography. British Patent 1315. [This patent adapts the Woodburytype process for pyrography.]

1866, July 24

WBW. An Improved Method of and Apparatus for Printing from Metal Intaglios (Obtained by the Aid of Photography) in Gelatinous or Other Semi-Transparent Ink. British Patent 1918. [This patent adapted the Woodburytype for the rotary press.]

1866, September 28

Williams, George F. "A Visit to Woodbury's Photographic Printing Establishment." *BJP* 13:465.

1866, November 9

"Talk in the Studio - Mr. Woodbury's Photo-Relief Printing Process." *PN* 10:540.

1866, December 28

"Producing Designs on Wood." *BJP* 12:620.

1867

WBW. "On a Method of Obtaining Imperishable Photographs." *British Journal Photographic Almanac* and *Photographer's Daily Companion* 1867: 95-6.

1867, March 30

WBW and Robert Ashton. Means of Producing Designs upon Paper. British Patent 947 [This patent adapts the Woodburytype process for watermarking. It was known as photo-filigrane.]

1868, April 17

"Mr. Woodbury's Photo-Relief Process." *PN* 12:132.

1868, June 5

"Echoes of the Month: Mr. Woodbury's Process." *PN* 12:266.

1869

Robinson, H. P. *Pictorial Effect in Photography*. London: Piper & Carter. [Contains a Woodburytype frontispiece.]

1869, March 16

WBW. "Description of the Woodburytype or Photo Relief Process." *Photographic Journal*: 4-6. [Probably the first time the processes was called by the name *Woodburytype*.]

1869, September 24

"Talk in the Studio - Mr. Woodbury's Photo-Relief Process." *PN* 13:465.

1869, November 12

"Goupil and Co.'s Woodburytype Establishment." *BJP* 16:544.

1869, November 12

Pritchard, H. Baden. "The Photo-Relief Process in Paris." *PN* 13:542-3.

1869, December

Minutes from the meeting of the Philadelphia Photographic Society. *PP* 6:422. [Edward Wilson exhibits first Woodburytype prints in America.]

1870

The *Photographic Art Journal*. London: Sampson Low, Son, and Marston.

1870, January

"The Woodbury Photo-Relief Printing Process—Our Picture," with frontispiece *PP* 7:2.

1870, January

Letter to the editor from John Carbutt. *PP* 7:32.

1870, February 7

WBW. A "method of producing surfaces by the aid of photography." British Patent 346. [This patent adapts the Woodburytype process for printing with traditional printers' ink using a grain in the matrix.]

1870, March

Minutes from the meeting of the Photographic Society of Philadelphia. *PP* 7:77-8 [First Woodburytype print exhibited by Carbutt.]

1870, April

"The Woodbury Process in France." *PP* 7:137-9.

1870, August

Simpson, George Wharton. "Notes in and Out of the Studio." *PP* 7:301-2.

1870, August 26

Pritchard, H. Baden. "The Woodbury and Lichtdruck Processes Compared." *PN* 14:398-9.

1870, September

Simpson, George Wharton. "Notes in and Out of the Studio." *PP* 7:318-21.

1870, October

———. "Notes in and Out of the Studio." *PP* 7:359-62.

1871

Wynter, Andrew. *Peeps into the Human Hive*. Vol. 1. London: Chapman and Hall.

1871, January

"Society for the Advancement of Photography, Berlin." *Anthony's Photographic Bulletin*: 19. Reprinted from *BJP*.

1871, August 25

Wortley, H. Stuart. "Official Report on Photography in the International Exhibition, 1871." *BJP* 18:400-1.

1872, August 16

"The Demise of the Woodbury Patent." *BJP* 19:383-5.

1872, November 15

Sutton, Thomas. "Woodbury's Process as Employed by MM Goupil at Paris." *BJP* 19:543-5.

1872, November 15

———. "The Woodburytype—Comparison Between it and the Heliotype—[Suggested use of the Latter in Portraiture]." *BJP* 19:548-9.

1872, December 4

WBW. Improvements in Photo-Mechanical Printing, and in Apparatus and Appliances Therefore. British Patent 3654. [This patent covers 1) making half-tone photographic type blocks using netting for printing with traditional ink 2) amendments to the Woodburytype process - the first hint of the Stannotype process and 3) adapting the Woodburytype for rotary printing.]

1873

The *Photographer's Friend*: 30.

1874

Tissandier, Gaston. La Photoglyptie. Paris: Dunod.

1874, July 10

WBW. "Photo Block Printing." PN 18:329.

1874, September

WBW. "Science at Home." *The Magic Lantern*: 13.

1885, September 11

"The Late Walter Bentley Woodbury." *BJP* 32:581.

1885, September 11

"Death of Walter Bentley Woodbury." *PN* 29:578-9.

1885, September 18

"The Late W.B. Woodbury." *BJP* 32:596.

1885, September 18

"Notes." *PN* 29:600. [Funeral notice]

1885, September 25

"Editorial Mirror." *PT* 15:546.

1885, October

"Obituary: Walter Bentley Woodbury." *PP* 22:340-1.

1885 November

"Walter Bentley Woodbury." *PP* 22:345-8.

1885, November

"A Worthy Life Ended: Walter Bentley Woodbury." *PP* 22:368-70.

1887

Advertisement for "The Woodbury Dry Plate" manufactured by William Grisom, Jr. Thos. H. McCollin & Co. trade agents. Philadelphia. *American Journal of Photography*.

1887

Burton, WK. *Practical Guide to Photographic & Photo-Mechanical Printing*. London: Marion & Co.

1889, September 13

Bolas, Thomas. "The Photo-Mechanical Printing Methods as Employed in the Jubilee Year of Photography." *BJP* 36:600-2.

1890

Chandler, Charles. "Photo-Mechanical Processes." in *Exhibition of Pictures By*

Modern Mechanical Photographic Processes, November 13-15. New York: New York Camera Club.

1890

Waterhouse, J. *Practical Notes on the Preparation of Drawings for Photographic Reproduction*. London: Kegan Paul, Trench, Trubner & Co.

1892

"Photo-Mechanical Processes." in The Reproductive Arts. Exhibition Catalogue. Museum of Fine Arts Boston.

1893

Robottom, Arthur. *Travels in Search of New Trade Products*. London: Jarrold & Sons.

1893, December 15

"The Fathers of Photography IV: Walter B. Woodbury." *PT* 23:734-5.

1894,

Woodbury, Walter Edward. "Photo-Mechanical Printing Processes." preface to *Catalogue of the Exhibition of Photo-Mechanical Prints and Printing Processes*. New York: Society of Amateur Photographers.

1894, November 16

"Early Half-Tone Cuts." *PT* 24:313.

1898

Woodbury, Walter Edward. *The Encyclopaedic Dictionary of Photography*. New York: Scoville & Adams. Reprint 1979. Arno Press.

ca. 1902

Exhibition Catalogue, Woodbury Gallery. London: Eyre & Spotiswoode.

1905 September

"Obituary: Walter Edward Woodbury." *Photo Miniature* 6:667.

1929

Swan, Mary E. and Kenneth R. Swan. *Sir Joseph Wilson Swan, F.R.S., Inventor & Scientist*. Newcastle upon Tyne: Oriel.

1935

Ferguson, Ellen Woodbury. "What I know about the Woodbury & Bentley Families." Unpublished family history. Collection of Luanne Woodbury.

1935

———. "Walter Bentley Woodbury." Unpublished biography. Collection of Luanne Woodbury.

1945

Eder, Joseph Maria. *History of Photography*. Translated by Edward Epstean. New York: Columbia University Press.

1957

Mertle, J. S. "The Woodburytype: The Most Beautiful Photographic Reproduction Process Ever Invented." *Image* 6:165-70.

1957

———. "Evolution of the Rotogravure." Private printing. Reprinted from *Gravure Magazine*. Nov 1955 to June 1957.

1965

Hall, Norman. "The Woodburytype." in *BJP* Annual. London: Henry Greenwood & Company.

1969

Gernsheim, Helmut and Alison Gernsheim. *The History of Photography from the Camera Obscura to the Beginning of the Modern Era*. 2nd ed. London: Thames & Hudson.

1971

Langford, M. J. "Sensitometry." Chap. 13 in Basic Photography, *A Primer for Professionals*. 2nd ed. New York: Focal Press.

1977

Steinberg, S. H. *500 Years of Printing*. With a foreword by Beatrice Ward. New York: Penguin.

1978

Byatt, Anthony. *Picture Postcards and Their Publishers*. Malvern, Worcs: Golden Age Postcard Books.

1978

Welling, William. *Photography in America: The Formative Years, 1839-1900*. Albuquerque: University of New Mexico Press.

1979

Crawford, William. *Keepers of Light*. New York: Morgan and Morgan.

1984

Brey, William. *John Carbutt on the Frontiers of Photography*. Cherry Hill, New Jersey: Willowdale Press.

1986

White, Steven. *John Thomson: A Window into the Orient*. London: Thames & Hudson.

1989

Hammond, Anne Kelsey. "Aesthetic Aspects of the Photomechanical Print." in *British Photography in the Nineteenth Century: The Fine Art Tradition*. Mike Weaver, ed. Cambridge University Press.

1991

Barger, M. Susan and William B. White. *The Daguerreotype: Nineteenth Century Technology and Modern Science*. Baltimore: Johns Hopkins University.

1994

Wachlin, Steven. *Woodbury & Page: Photographers Java*. Leiden: KITLV.

No date

Woodbury, Luanne. Unpublished Woodbury family tree. Collection of Luanne Woodbury.

NOTES

Notes

Abbreviations used in the notes:

BJP—British Journal of Photography

NMPFT—National Museum of Photography, Film and Television, Bradford, England

PP—Philadelphia Photographer

PN—Photographic News

RPS—Royal Photographic Society

WBW—Walter Bentley Woodbury

WBT—Woodburytype

Chapter 1: Walter Bentley Woodbury—The Early Years

1 Unpublished family history written by Ellen Woodbury Ferguson, in the collection of Luanne Woodbury, London. And "The Fathers of Photography: IV—Walter Bentley Woodbury," *Photographic Times and American Photographer.* (1893): 734-5. This unsigned editorial piece was probably written by Walter E. Woodbury, Woodbury's son, who was associate editor of the *Photographic Times.*

2 Unpublished family tree written by Luanne Woodbury, London.

3 Unpublished biography of Walter Bentley Woodbury, written by Ellen Woodbury Ferguson, in the collection of Luanne Woodbury, London. Hereafter referred to as WBW Bio.

4 *Philadelphia Photographer* 22 (October 1885): 345.

5 "At Home: Mr. Walter Woodbury at Manor House, South Norwood," *Anthony's Photographic Bulletin* (1881): 72-74.

6 Unpublished letter from WBW to his mother, April 24, 1853. (In the Royal Photographic Society collection at the National Museum of Photography, Film and Television, Bradford, England; hereafter referred to as RPS Coll.) And Steven Wachlin, Woodbury & Page (Leiden: KITLV press. 1994), 9.

7 WBW to his mother, November 21, 1852 (RPS Coll.).

8 Ibid

[64] W. K. Burton, *Practical Guide to Photographic & Photo-Mechanical Printing* (London: Marion & Co. 1887), 15.

[65] William Crawford expands on Ivins's theories, developing the idea of camera syntax—the visual peculiarities associated with choice of camera, lens and film. This should be regarded as separate from any syntax imposed on an original image by the reproduction medium. William Crawford, *Keepers of Light* (New York: Morgan and Morgan, 1979).

[66] "A New System of Printing," *PN* 8 (October 14, 1864): 493-494.

[67] Ibid.

[68] John Thomson, editorial note in Gaston Tissandier, *History and Handbook of Photography*, translated and edited by John Thomson (London: Sampson Low. 1876), 205.

Chapter 4: The Woodburytype and the Birth of Visual Democracy

[69] *BJP* 12 (October 20, 1865): 535.

[70] WBW, ed. Preface, *Treasure Spots of the World: A Selection of the Chief Beauties and Wonders of Nature and Art* (London: Ward, Lock, and Tyler, 1875).

[71] John Thomson and Adolphe Smith, *Street Life in London* (London: Sampson Low, Marston, Searle and Rivington, 1877-1878).

[72] This is not the only book for which Thomson used the Woodburytype. See Steven White, *John Thomson: A Window to the Orient* (Thames & Hudson, 1986).

[73] Adolphe Smith, "The 'Crawlers'," in *Street Life*.

[74] We know, however, that these books were not available to all people. The working classes still could not afford a book of this type. But as the middle classes were growing the apparency of open, democratic government began to rise. See "The Woodbury Permanent Printing Company at Kent Gardens, Ealing," *PN* 24 (March 25, 1880).

[75] This transition is evident in a similar work published by John Carbutt, owner of the Woodburytype rights in the United States. In 1869, before Carbutt started working the Woodburytype, he released the publication *Biographical Sketches of the Leading Men of Chicago, Photographically Illustrated.* The book contained photographic portraits of some one-hundred people, printed with real silver

prints in an edition of about 500, bringing the total to a massive 50,000 photographs. William Welling, *Photography in America, The Formative Years, 1839-1900.* (Albuquerque: University of New Mexico Press), 202.

[76] Because many of the *Galerie Contemporaine* volumes have been pulled apart for their plates, and in many cases re-bound, identifying volume numbers is difficult.

[77] I believe this and similar publications helped raise enormous awareness of Paris for foreign artists and helped propel France into the forefront of early twentieth century art.

[78] Exhibition catalogue, Woodbury Gallery. (London: Eyre & Spottiswoode, c. 1902). Eyre & Spottiswoode, the successors to the Woodbury Permanent Printing Company ran a shop called the Woodbury Gallery which displayed original works of art and presumably reproductions printed by the firm.

[79] *Photographic Art Journal* (London: Sampson Low, Son, and Marston. 1870). This journal was continued under the name *Art Pictorial and Industrial.*

[80] *Photographic Art Journal*: 2.

[81] *PN* 10 (November 9, 1866): 540.

[82] Eder, *History of Photography*: 588.

[83] *Photographic Art Journal*: 2.

[84] The Firm M. M. Goupil et Cie. later became Valadon, Boussod et Cie and it was during this time, from 1890 on, that the Woodburytype's use declined.

[85] J. S. Mertle, *Evolution of the Rotogravure*, Part 7 (Private printing: 1957), 23. Reprinted from *Gravure Magazine*, November 1955 to June 1957.

[86] H. Baden Pritchard, "The Woodbury Process in France," *PP*. Excerpted from *Photographic Journal* 1870: 137.

[87] *BJP* 16 (November 12, 1869).

[88] Other plants working the Woodburytype, including the WPPPC in England, also installed electric lights.

[89] Unidentified article in Walter Woodbury's "Newspaper Cuttings" book in the RPS Collection, NMPFT. Hereafter referred to as Woodbury's Scrapbook.

[90] WBW, "Woodbury's New Printing Process," *American Journal of Photography*, (May 15, 1865): 515. Reprinted from *BJP* 12 (March 17, 1865). And *PP* 3 (January 1866): 20. And the Minutes of the Philadelphia Photographic Society for

Wednesday November 3, 1869, *PP* 6 (December 1869): 422.

[91] William Brey, *John Carbutt on the Frontiers of Photography* (Cherry Hill, New Jersey: Willowdale Press, 1984), 96.

[92] Museum of Fine Arts Boston, Exhibition Catalogue, The Reproductive Arts, 1892: 63. And *PP* 7 (November 1885): 369.

[93] *PP* November 1885: 369. A note of interest here that Carbutt left his Chicago facilities in 1870 and they were almost immediately destroyed by fire. William Brey, *John Carbutt.*

[94] "The Demise of the Woodbury Patent," *BJP* 19 (August 16, 1872): 383.

[95] "Goupil and Co's Woodburytype Establishment," *BJP* 16 (November 12, 1869): 544-545.

[96] *PN* 12 (April 17, 1868): 182.

[97] "Echoes of the Month," *PN* 12 (June 5, 1868): 267.

[98] G. W. Simpson, *PP* 7 (August 1870): 302.

[99] Anthony Byatt, *Picture Postcards and Their Publishers* (Malvern, Worcs: Golden Age Postcard Books. 1978), 92.

[100] G. W. Simpson, "Notes in and out of the Studio," *PP* 7 (August 1870): 321.

[101] Indent of Materials for the Woodbury Process, Government printing office, Sydney, June 29, 1878. Document in the collection of Luanne Woodbury, London.

[102] Conversation between Michael Gray and the author, November 14, 2004. Waterhouse was an advocate of the Woodburytype and described its use in several lectures, including "The Application of Photography to the Reproduction of Maps and Plans by Photo-Mechanical and other processes," presented to the Geographical Congress at Paris 1875, printed in *Journal of the Asiatic Society of Bengal,* 1878.

[103] Minutes for Society for the Advancement of Photography, Berlin, January 1871, *Anthony's Photographic Bulletin* (1871): 21

[104] In 1869 Woodbury claimed his working plates were pulling around 1000 prints each and that he could print up to 40,000 prints per week. The quantity was really limited only by time. It is possible to produce an unlimited number of matrices from a single negative, producing more each time one wears out. George F. Williams, "A Visit to Woodbury's Photographic Printing Establishment,"

BJP 13 (September 28, 1866): 465-467. And WBW, "A Description of the Woodburytype Process," *Photographic Journal* (March 16, 1869): 4-6.

[105]*PN* 13 (September 24, 1869): 465. And *Photographic World* (January 1871): 19. Comparing these figures to the edition size of 500 for Carbutt's *Leading Men of Chicago* shows the progress that had been made.

[106]George F. Williams, "A Visit to Woodbury's Photographic Printing Establishment."

[107]Andrew Wynter, *Peeps into the Human Hive*, Volume I (London: Chapman and Hall, 1871), 61.

[108]*PN* 10 (January 26, 1866): 38.

[109]Helmut and Alison Gernsheim. *History of Photography* (London: Thames and Hudson. 1969), 544.

[110]"Review of the Progress of Photography during the Year 1869," *Photographic Art Journal* (1870): 5. A notice for Woodburytype slides sold at 75 cents each appears in *The Magic Lantern* 8 (May 1874): 9.

[111]"Mr. Walter B. Woodbury," *PN* 29 (March 13, 1885): 165.

[112]Society for the Advancement of Photography, Berlin, January 1871.

Chapter 5: Technical Elements of the Woodburytype

[113]WBW, An improved method of producing or obtaining by the aid of photography, surfaces in "relievo" and "intaglio" upon aluminous, vitreous, metallic, or other suitable materials. British Patent 2338, September 23, 1864.

[114]The modern chemical name for this class of salts is now Dichromate (i.e. Potassium Dichromate).

[115]This should not be confused with the process developed by Talbot that he termed photoglyphic printing, an entirely different process.

[116]Eder, *History of Photography*: 552.

[117]Henry H. Snelling, *The History and Practice of the Art of Photography*, (New York: Putnam, 1849; Repr., Morgan & Morgan, 1970), 112.

[118]Eder, *History of Photography*: 552-555.

[119]There are a number of books and articles that give detailed formulae and procedure for producing Woodburytypes. *The Encyclopaedic Dictionary*, by Walter E. Woodbury probably contains the best description.

[120] An actinometer is an apparatus for measuring actinic (or photographically active) light. It was used to judge the cumulative amount of exposure a sensitive film had received.

[121] As related by John Carbutt of the American Photo-Relief Printing Company in *The Photographer's Friend* (1873): 30.

[122] The International Museum of Photography at the George Eastman House has many fine examples of Woodburytype slides printed by the American Photo-Relief Printing Company.

[123] Eder, *History of Photography*: 588. And the Carbutt collection at the George Eastman House.

[124] S. H. Steinberg, 500 Years of Printing (New York: Penguin, 1977), 277-291. This book gives an excellent description of the technical advancements in literary printing at the time, including changes in paper manufacture, letterpress, typesetting and illustration.

[125] WBW, An improved method of and apparatus for printing from metal intaglios (obtained by the aid of photography) in gelatinous or other semi-transparent ink. British Patent 1918, 24 July 1866. And George F. Williams, "A Visit to Woodbury's Photographic Printing Establishment."

[126] One imperfect although quite beautiful experiment of a rotary Woodburytype exists in the RPS collection, but no account of commercial use exists. See plate 49.

[127] This should not be confused with the goupil-gravure developed by Woodbury.

[128] Walter E. Woodbury, "Photo-Mechanical Printing Processes," Preface, Catalogue of the Exhibition of Photo-Mechanical Prints and Printing Processes (New York: Society of Amateur Photographers, 1894), 4.

[129] Mertle, *Evolution of the Rotogravure*: 23, gives an excellent description of the Goupil-gravure process.

[130] Roussellon continued to refine and perfect the process after Woodbury's visit and later claimed to be its inventor, applying for recognition to the French Scientific Society. These claims, like others that plagued Woodbury, eventually turned legal, but the outcome is uncertain. Woodbury's Scrapbook.

[131] Samples of this process were exhibited, along side Woodburytype prints, at the 1871 International Exhibition to favorable review. See H. Stuart Wortley, "Official

Report on Photography in the International Exhibition of 1871," *BJP* 18 (August 25, 1871): 400-401.

[132] WBW, "Photo Block Printing," *PN* 18 (July 10, 1874): 329.

[133] The Ives process, the first practical color halftone process, was originally based in part on the Woodburytype. He surface inked a Woodburytype plate and pressed it into contact with a textured paper to produce the dots he needed to separate the tones of the image. Ives's U.S. patent of August 12, 1878 reads, "I, the undersigned, have today invented a method of obtaining relief plates for the typographic printing press from ordinary photographic negatives, which may be described as follows: —1st, From an ordinary photographic negative, a relief in gelatin, similar to that used in the Woodburytype process, but perhaps in lower relief, is obtained." Thomas Bolas, *BJP* 36 (September 13, 1889).

[134] "Photographs Transformed into Typographic Blocks," *PN* 27 (October 26, 1883): 678.

[135] WBW and Robert Howe Ashton, Means of producing designs on paper. British Patent 947, March 30, 1867.

[136] *The Yorkshireman* (1883): 76.

[137] WBW, Improvements in producing designs upon wood and other materials by the aid of photography. British Patent 1315, May 8, 1866. And "Producing Designs on Wood," *BJP* 13 (December 28, 1866): 620.

[138] "The Stannotype Process," *BJP* 29 (February 17, 1882): 88-89.

[139] The halftone, or screen process continues to this day as the chief means for printing photographic images by press and now digital technology.

[140] Advances in paper manufacturing technology also played a part in this. New papers, able to hold finer patterns of ink without bleeding, made possible the development of dot-based images like the halftone.

[141] Goupil & Co, *Exhibition of Pictures By Modern Mechanical Photographic Processes* (New York: New York Camera Club 1890), 29.

[142] This process was also perfected by several other inventors and known by various names, such as the Heliotype of Ernest Edwards, and the Lichtdruck process.

[143] Society for the Advancement of Photography, Berlin, January 1871.

[144] J. Waterhouse, *Practical Notes on the Preparation of Drawings for Photographic*

Reproduction (London: Kegan Paul, Trench, Trubner & Co. 1890), 33.

Chapter 6: Evolution of the Woodburytype Process And Woodbury's Later Years

[145]WBW, "Woodbury's New Printing Process," *BJP* 12 (March 17, 1865): 134.

[146]Editor's note in WBW, "Woodbury's New Printing Process," *BJP* 12 (March 17, 1865).

[147]This is not the only invention for which Woodbury had competing claims. Others included the Goupil-gravure, the Stannotype, balloon photography and stripping negatives from glass. See note 130.

[148]Commonly called carbon tissue, this is a misnomer and is actually a thin sheet of gelatin film, usually coated on high-grade paper rather than a tissue.

[149]WBW, "Correspondence: The 'Woodburytype' Process," *PN* 29 (July 24, 1885): 477. Although in the *PN* October 13, 1865 article, Woodbury states, "little more than twelve months have elapsed since my first experiments on the subject (having had no previous experience with the materials used)." It is quite likely that since the later article is a recollection that some exaggeration was used.

[150]W. K. Burton, *Practical Guide to Photographic & Photo-Mechanical Printing* (London: Marion & Co. 1887), 13.

[151]This person was probably George Wharton Simpson, who both gentlemen refer to in their correspondence. See later in this chapter for more details.

[152]Joseph Wilson Swan, letter to the editors "Woodbury's New Printing Process," *BJP* 12 (April 7, 1865): 182.

[153]"A New Carbon Process," *PN* 8 (February 19, 1864): 85.

[154]Swan, *BJP* 12 (May 5, 1865): 238.

[155]Ibid.

[156]Swan, *PN* 9 (October 20, 1865): 502.

[157]WBW, *PN* 9 (October 13, 1865): 490.

[158]"Talk in the Studio, The Photo-Relief Printing Process," *PN* 10 (April 27, 1866): 203.

[159]WBW, "Description of the Woodburytype or Photo Relief Process," *Photographic Journal* (March 16, 1869): 4-6. Probably the first time the process was called by the name *Woodburytype*.

[160]*PN* 10 (January 26, 1866): 38.

[161] George Wallis, New or improved methods of preparing drawings, writings, designs, prints, or impressions of engravings and photographs, for the purpose of impressing or engraving the same in or upon metallic surfaces, and thereby producing printing or embossing surfaces or ornamental metallic surfaces for such purposes as the same are or may be applicable to; also new or improved machinery to be employed in the said impressing or engraving. British Patent 1299, May 25, 1860.

[162] Wallis gave a detailed account of the history of nature printing in his paper "The New Art of Auto-Typography," (see note 165 for details). According to Wallis's history all the developments had been made during the nineteenth century. Eder, in his *History of Photography* gives the credit to Alois Auer of Vienna. Wallis mentions the government printing office where Auer worked but does not cite him as a source. Many history text have used Eder's story and Wallis's, written first, gives a different interpretation.

[163] Eder, *History of Photography*: 33-34.

[164] Minutes from the meeting of the South London Photographic Society, *PN* 7 (May 22, 1863): 248-250.

[165] George Wallis, "The New Art of Auto-Typography," *Society of Arts Journal* (April 17, 1963) reprinted *PN* 7 (April 24, 1963).

[166] "Mr. Swan's Process and the Printing Press," *PN* 8 (May 20, 1864): 242.

[167] This article was unsigned, but probably written by Simpson, editor of *PN*, who was present for Wallis's presentation.

[168] This point came up for discussion when Wallis presented his process to the South London Photographic Society.

[169] See Woodbury's experiments in this direction in the previous chapter.

[170] Woodbury attributed this discovery in 1869 to Mr. H. Cooper. See WBW, "Description of the Woodburytype, or Photo-relief Process," *Photographic Journal* (March 16, 1869): 5.

[171] W. K. Burton, *Practical Guide to Photographic & Photo-Mechanical Printing*: 14.

[172] *PN* 9 (August 18, 1865): 387.

[173] Woodbury was adamant about and repeated often the importance of having a completely flat, equal and planed surface on both sides at all steps in the

process. Even a fraction of an inch difference would produce unevenness in the final prints.

[174] Swan, Improvements in the production of printing surfaces by photographic agency, and in obtaining prints therefrom. British Patent No. 1791, July 6, 1865.

[175] This firm was Disderi & Company, of England. Eder, *History of Photography*: 588.

[176] *BJP* 19 (August 16, 1872): 383.

[177] Gernsheim, *History of Photography*, 340.

[178] The most extensive treatise on the Woodburytype to date was written based on the new modifications of the Stannotype. Leon Vidal, *Traite Practique de Photoglyptie* (Paris: Gautheir-Villars, 1881).

[179] Unfortunately I have not been able to locate any prints labeled as Stannotypes.

[180] "At Home: Mr. Walter Woodbury at Manor House, South Norwood," *Anthony's Photographic Bulletin* (1881): 72-74.

[181] *BJP* 27 (May 21, 1880): viii. Advertisement for carbon enlargements printed by the Woodbury Permanent Photographic Printing Company on their specially made carbon tissue, which was developed for the Woodburytype process. And *American Journal of Photography* (1887), Advertisement for Woodbury Dry Plates, manufactured by the Woodbury Dry Plate Company.

[182] *PP* 22 (May 1885): 144.

[183] Unidentified article in Woodbury's Scrapbook.

[184] Leon Vidal, *Agenda Photographique* (Paris: Librairie du Monituer Universel, 1876), 83. And "Mr. W. B. Woodbury," *BJP* 31 (March 14, 1884): 166.

[185] Unidentified article in Woodbury's Scrapbook. It is most likely from one of the British photographic journals, dated between 1876 and 1884.

[186] Woodbury wrote more articles on balloon photography than any of his other non-Woodburytype inventions. Quite a few appeared in the popular press. See "Walter B. Woodbury," *PN* 29 (March 13, 1885): 165-166.

[187] "Walter B. Woodbury," *PN* 29 (March 13, 1885): 165-166. And WBW, "Science at Home: I," The magic Lantern 9 (May 1875): 13.

[188] WBW, "The 'Woodburytype' Process," *PN* 29 (July 24, 1885): 447. In this article Woodbury attempts to "correct the errors and misstatements which have somehow crept into," a placard, displayed by the Woodbury Permanent Printing Company

at an exhibition, that gave most of the credit to Swan for the discoveries of the process.

[189]In 1875, *Treasure Spots of the World* was published, edited by Woodbury with photographs by himself and others reproduced in Woodburytype. The book appears more as a promotional device for the process than for Woodbury as a photographer.

[190]*PN* 29 (July 10, 1885): 433-434.

[191]*PP* 22 (November 1885): 370.

[192]*PP* 22 (October 1885): 341.

[193]*BJP* 32 (September 11, 1885): 581.

[194]"The Death of Walter Bentley Woodbury," *PN* 29 (September 11, 1885): 578-579.

[195]Ibid.

[196]*PP* 22 (October 1885): 348.

[197]*PP* 22 (October 1885): 340-341.

[198]*PP* 22 (November 1885): 369.

[199]Walter E. Woodbury, *The Encyclopaedic Dictionary of Photography* (New York: Scovill & Adams, 1898; repr., Arno Press, 1979).

[200]*Photo Miniature* (September 1905): 667. Coincidentally, John Carbutt, the holder of the Woodburytype patent for America, died exactly one month after Walter E., his obituary appeared on the facing page of *Photo Miniature*.

[201]Eder, *History of Photography*: 5

INDEX

COLOPHON

This edition of A History of the Woodburytype *was published by Carl Mautz Publishing, Nevada City, California, 2007.*

The text and titles were composed in the Bauer Bodoni and Helvetica Neue types. Bauer Bodoni was released in 1926 and designed by Heinrich Jost. The forms are closely related to the original Bodoni typefaces designed by Giambattista Bodoni in the 1800s. Helvetica Neue was released in 1983 by Linotype type house, and is a reworking of Helvetica, the sans-serif typeface developed in 1957 by Max Miedinger.

Cover and book design, typography, and production were by Susan Powell, Nevada City, California.

Printed in Korea by Sung In Printing America, Inc.